Microsoft Certified Azure Data Fundamentals (DP-900) Exam Guide

Build a solid foundation in Azure data services and pass the DP-900 exam on your first try

Steve Miles

‹packt›

Microsoft Certified Azure Data Fundamentals (DP-900) Exam Guide

Author: Steve Miles

Reviewer: Priyanka Agrawal

Publishing Product Manager: Anindya Sil

Senior-Development Editor: Ketan Giri

Development Editor: Kalyani S.

Digital Editor: M Keerthi Nair

Presentation Designer: Shantanu Zagade

Editorial Board: Vijin Boricha, Megan Carlisle, Simon Cox, Ketan Giri, Saurabh Kadave, Alex Mazonowicz, Gandhali Raut, and Ankita Thakur

First Published: September 2024

Production Reference: 1260924

Published by Packt Publishing Ltd.
Grosvenor House
11 St Paul's Square
Birmingham
B3 1RB

ISBN: 978-1-83620-815-0

www.packtpub.com

Contributors

About the Author

Steve Miles works in a senior technology role for the cloud practice of a multi-billion turnover European IT distributor. He is a Microsoft Most Valuable Professional (MVP), Microsoft Certified Trainer (MCT), and an Alibaba Cloud MVP with 25+ years of technology experience in hosted datacenter services, hybrid, and multi-cloud platforms, and a previous military career in engineering, signals, and communications. Steve is the author of many books on Microsoft technologies with a focus on Azure, AI and data, as well as security, which can be found on his author profile on Amazon at `https://www.amazon.com/stores/Steve-Miles/author/B09NDJ1RC8`.

Steve is a petrolhead and can be found tinkering with cars when he is not writing.

You can connect with him on LinkedIn at `https://www.linkedin.com/in/stevemiles70/`

About the Reviewer

Priyanka Agrawal is a Technical Trainer at Microsoft, USA, with over 15 years of experience as a Microsoft Certified Trainer. She is passionate about learning and sharing knowledge in all capabilities and excels in delivering training, proctoring, and upskilling Microsoft Partners and Customers. Priyanka has significantly contributed to AI and Data-related courseware, exams, and high-profile events such as Microsoft Ignite, Microsoft Learn Live Shows, MCT Community AI Readiness, and Women in Cloud Skills Ready.

Beyond her professional achievements, Priyanka is a passionate advocate for environmental protection and actively supports related initiatives. In her personal time, she enjoys traveling and spending quality time with her family.

Table of Contents

3

4

5

6

7

8

9

Preface

The **DP-900 certification exam**, also called the **Azure Data Fundamentals exam**, is designed to validate foundational knowledge of data concepts and how they are implemented in Microsoft Azure.

The DP-900 exam has been updated a few times to include new technologies as they emerge and enter the Azure environment—and the latest edition of the exam is no different. The DP-900 exam now includes content on the Microsoft Fabric unified data analytics platform as part of an Azure subscription.

Who This Book Is For

This book is intended for individuals who wish to pass the Microsoft Certified: Azure Data Fundamentals (DP-900) exam, to demonstrate their knowledge and skills in this area, as well as those interested in gaining a basic understanding of Azure's data capabilities who might not have experience in the field. This includes business stakeholders, decision-makers, and technical professionals who are new to data and cloud technologies.

The content in this book assumes you have no knowledge of data concepts (though having that knowledge would certainly help with understanding the fundamental cloud computing topics covered).

Written in a clear, concise way and comprising self-assessment questions, exam tips, and mock exams with detailed answers, this book covers the topics of data concepts and Azure's data services and provides a focus on the exam skills measured, without burdening you with complex jargon or unnecessary background information or concepts, allowing you to pass the DP-900 exam with confidence.

What This Book Covers

Chapter 1, Describe Core Data Concepts, includes content covering core data concepts such as structured, semi-structured, and unstructured data features; common formats for data files and types of databases; the features of transactional and analytical workloads; and the responsibilities of data-related roles such as administrators, engineers, and analysts.

Chapter 2, Describe Relational Concepts, includes content covering the features of relational data, understanding normalization, and identifying common **Structured Query Language** (**SQL**) statements and common database objects.

Chapter 3, Describe Relational Azure Data Services, includes content covering the SQL product family available in Azure and open source database systems.

Chapter 4, Describe the Capabilities of Azure Storage, includes content covering the Blob, File, and Table Storage services available in Azure.

Chapter 5, Describe the Capabilities and Features of Azure Cosmos DB, includes content covering Azure Cosmos DB use cases and the APIs of Azure Cosmos DB.

Chapter 6, Describe the Common Elements of Large-Scale Analytics, includes content covering data ingestion and processing considerations, analytical data store options, and data warehousing services in Azure.

Chapter 7, Describe Consideration for Real-Time Data Analytics, includes content covering the differences between batch and streaming data and exploring real-time analytics services in Azure.

Chapter 8, Describe Data Visualization in Microsoft BI, includes content covering Power BI capabilities, data model features, and data visualizations.

How to Get the Most Out of This Book

This book is crafted to equip you with the knowledge and skills necessary to excel in the DP-900 exam through memorable explanations of major domain topics. It covers the eight core domains critical to the security expertise that candidates must be proficient in to pass the exam. For each domain, you'll work through content that reflects real-world IT security challenges. At certain points in the book, you'll assess your understanding by taking chapter-specific quizzes. This not only prepares you for the DP-900 exam but also allows you to dive deeper into a topic as needed, based on your results.

Online Practice Resources

With this book, you will unlock unlimited access to our online exam-prep platform (*Figure 0.1*). This is your place to practice everything you learn in the book. How to access the resources. To learn how to access the online resources, refer to *Chapter 9, Accessing the Online Practice Resources* at the end of this book.

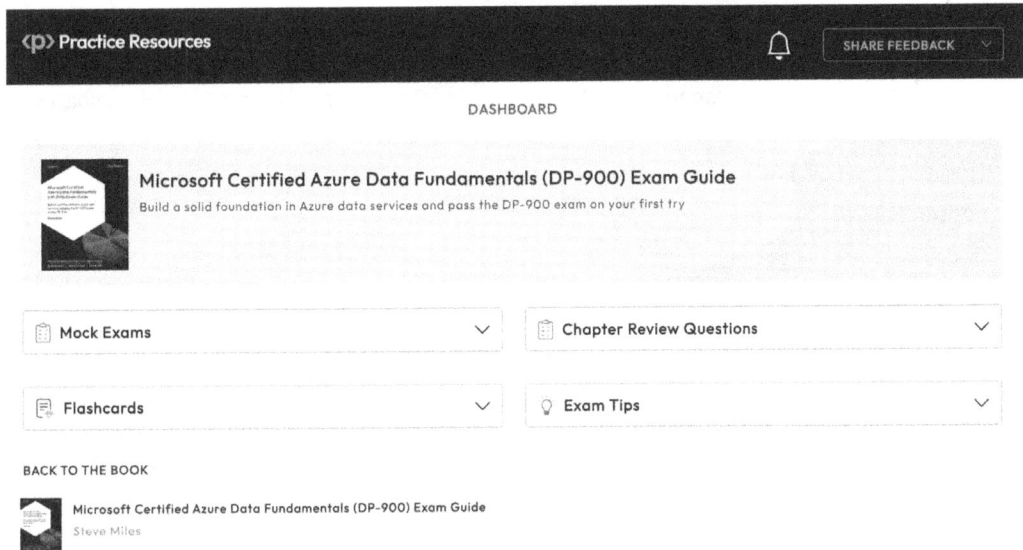

Figure 0.1 – Online exam-prep platform on a desktop device

Sharpen your knowledge of DP-900 concepts with multiple sets of mock exams, interactive flashcards, and exam tips accessible from all modern web browsers.

Download the Color Images

We also provide a PDF file that has color images of the screenshots/diagrams used in this book. You can download it here: `https://packt.link/DP900GraphicBundle`.

Conventions Used

`Code in text`: Indicates code words in the text, database table names, folder names, filenames, file extensions, pathnames, dummy URLs, user input, and X (formerly known as Twitter) handles are shown as follows: "The elements are enclosed within `<element>` and `</element>` tags and are hierarchical; an element can contain other elements, attributes, or text."

A block of code is set as follows:

```
SELECT *
FROM Employees
WHERE Department = 'Engineering';
SELECT ProductID, Name, Price
FROM Product
ORDER BY Price ASC;
```

Bold: Indicates a new term or an important word and abbreviations. Here is an example: "Data can be retrieved from across tables with **Structured Query Language (SQL)** statements; in brief, SQL is a standardized programming language used for managing and manipulating relational databases."

> **Tips or important notes**
> Appear like this.

Get in Touch

Feedback from our readers is always welcome.

General feedback: If you have any questions about this book, please mention the book title in the subject of your message and email us at customercare@packt.com.

Errata: Although we have taken every care to ensure the accuracy of our content, mistakes do happen. If you have found a mistake in this book, we would be grateful if you could report this to us. Please visit www.packtpub.com/support/errata and complete the form. We ensure that all valid errata are promptly updated in the GitHub repository at: https://packt.link/DP900repo.

Piracy: If you come across any illegal copies of our works in any form on the internet, we would be grateful if you could provide us with the location address or website name. Please contact us at copyright@packt.com with a link to the material.

If you are interested in becoming an author: If there is a topic that you have expertise in and you are interested in either writing or contributing to a book, please visit authors.packtpub.com.

Share Your Thoughts

Once you've read *Microsoft Certified Azure Data Fundamentals (DP-900) Exam Guide*, we'd love to hear your thoughts! Scan the QR code below to go straight to the Amazon review page for this book and share your feedback.

https://packt.link/r/1836208154

Your review is important to us and the tech community and will help us make sure we're delivering excellent quality content.

Download a Free PDF Copy of This Book

Thanks for purchasing this book!

Do you like to read on the go but are unable to carry your print books everywhere?

Is your eBook purchase not compatible with the device of your choice?

Don't worry, now with every Packt book you get a DRM-free PDF version of that book at no cost.

Read anywhere, any place, on any device. Search, copy, and paste code from your favorite technical books directly into your application.

The perks don't stop there, you can get exclusive access to discounts, newsletters, and great free content in your inbox daily.

Follow these simple steps to get the benefits:

1. Scan the QR code or visit the link below:

https://packt.link/free-ebook/9781836208150

2. Submit your proof of purchase.
3. That's it! We'll send your free PDF and other benefits to your email directly.

1

Describe Core Data Concepts

In this first chapter of this *DP-900 Microsoft Azure Data Fundamentals* exam guide, you will delve into the core data concepts that underpin these transformative trends. You will explore the fundamental principles that govern the structure, processing, analysis, utilization, and management of data. In the second part, you will move on to understanding the evolving job roles and responsibilities required within modern data-driven organizations.

This chapter primarily focuses on the *Describe Core Data Concepts* module from the *Skills Measured* section of the *DP-900 Microsoft Azure Data Fundamentals* exam.

This chapter's content requires no prior understanding of data and does not necessarily require you to be in an existing data or technical role to learn the concepts and pick up the skills from this chapter.

Making the Most Out of This Book – Your Certification and Beyond

This book and its accompanying online resources are designed to be a complete preparation tool for your **DP-900 Exam**.

The book is written in a way that you can apply everything you've learned here even after your certification. The online practice resources that come with this book (*Figure 1.1*) are designed to improve your test-taking skills. They are loaded with timed mock exams, interactive flashcards, and exam tips to help you work on your exam readiness from now till your test day.

> **Before You Proceed**
>
> To learn how to access these resources, head over to *Chapter 9, Accessing the Online Practice Resources*, at the end of the book.

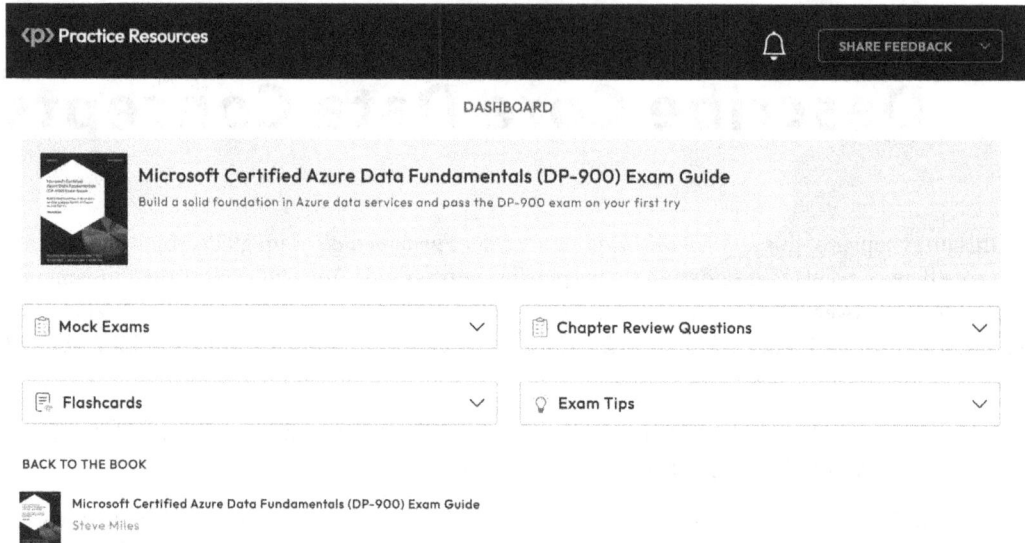

Figure 1.1: Dashboard interface of the online practice resources

Here are some tips on how to make the most out of this book so that you can clear your certification and retain your knowledge beyond your exam:

1. Read each section thoroughly.

2. **Make ample notes**: You can use your favorite online note-taking tool or use a physical notebook. The free online resources also give you access to an online version of this book. Click the BACK TO THE BOOK link from the Dashboard to access the book in **Packt Reader**. You can highlight specific sections of the book there.

3. **Chapter Review Questions**: At the end of this chapter, you'll find a link to review questions for this chapter. These are designed to test your knowledge of the chapter. Aim to score at least **75%** before moving on to the next chapter. You'll find detailed instructions on how to make the most of these questions at the end of this chapter in the *Exam Readiness Drill - Chapter Review Questions* section. That way, you're improving your exam-taking skills after each chapter, rather than at the end.

4. **Flashcards**: After you've gone through the book and scored **75%** more in each of the chapter review questions, start reviewing the online flashcards. They will help you memorize key concepts.

5. **Mock Exams**: Solve the mock exams that come with the book till your exam day. If you get some answers wrong, go back to the book and revisit the concepts you're weak in.

6. **Exam Tips**: Review these from time to time to improve your exam readiness even further.

By the end of this chapter, you will be able to answer exam questions on the following topics with confidence:

- Features of structured, semi-structured, and unstructured data
- Common data file formats
- Types of databases
- Transactional and analytical workloads
- Responsibilities for the roles of database administrators, data engineers, and data analysts

Your knowledge will be tested at the end of this chapter, and some questions will be asked to review your understanding of the topic.

In addition, this chapter's goal is to take your knowledge beyond the exam content so you are prepared for a real-world, day-to-day Azure data-focused role.

But before you continue with the content of this chapter, take a moment to pause and consider what it means to live in an era in which rapidly evolving technologies such as **artificial intelligence (AI)** are changing the world around you in real time. How are organizations going about solving their biggest challenges using data? What trends are transforming the world of work and changing the professional roles, tasks, and jobs you do every day?

Describe Ways to Represent Data

This book's first content section introduces you to data concepts and terminology. We will look at what data is, its definition, and how and why it is structured differently. Data structures are specialized formats for organizing, processing, retrieving, and storing data. They are essential for efficient data management and manipulation in computing.

What is Data?

Data is essentially information. When we are talking about modern IT systems, this data is stored as values and can be anything from a person's age stored in a database to a two-hour film stored in the servers of an on-demand video-streaming service.

All data starts as "raw data," collected from different sources without order and classification. Generally, this data needs to be filtered, sorted, and analyzed before it can be used. Raw data can be likened to crude oil; its potential and value cannot be derived when raw without action being taken to transform it into another form that can then be consumed in some more meaningful and valuable way. It takes a lot of resources and processes to turn something raw into something we can use that has value; otherwise, we just store the latent value that remains unlocked and unutilized. Gasoline is a good example of this transformation process. Crude oil is turned into gasoline, but not instantly; the refining process is the most crucial and critical. The same happens with raw data.

"Processed" data is data that has been transformed in some way, filtered, cleansed, organized, formatted, and analyzed to extract valuable information. Sales figures recorded daily would be raw data; in contrast, a report summarizing the sales trends during a monthly period would be the processed data. The processed data can then be stored in databases for further consumption for further derived value.

Data analysis refers to the steps of looking at numerical and other information, preparing and transforming it, and then modeling it to identify useful information, reach conclusions, and make decisions. Through this process, organizations may identify patterns, correlations, and trends from data that can inform strategic planning and operations. For example, by examining customer data, an organization can identify buying patterns and preferences that inform marketing efforts, the focus of their attention, and how they can best serve their customers.

Data Structures

Data values are typically organized into data **entities** and **attributes**.

These entities and attributes are central concepts in organizing data. A data entity represents a distinct object or concept about which data is stored. In a **customer relationship management (CRM)** system, for example, Customer is a data entity consisting of all the information about this particular customer, such as the customer ID, name, and email.

Each entity comprises many attributes (also known as **properties** or **fields**), which are specific features or facets of an entity. The attribute is a specific property or fact that concerns an entity; for example, `Customer` might have `CustomerID`, `Name`, `Mail`, and `PurchaseHistory` attributes.

In another example of a university scenario, entities might have the names `Students`, `Courses`, and `Professors`. Each of these entities has attributes that describe the characteristics of the entity they refer to. The `Students` entity may have the `StudentID`, `Name`, `Major`, and `EnrollmentDate` attributes. These attributes describe a student completely and uniquely, making entities of this kind distinguishable from one another.

Entities and attributes provide a structured way of organizing and making sense of data, facilitating its collection, analysis, and application.

Data Dimensions

When talking about data, you should consider that it has the following three **V** dimensions:

- **Volume**: This is the amount of data generated.
- **Velocity**: This is the speed at which data is generated.
- **Variety**: This is the diversity of data generated.

Data volume refers to the amount of data that is produced and captured. As AI technologies improve, data volume goes up since more data means better chances of training and developing more accurate models.

For instance, a self-driving car must gather thousands of examples from sensor data and camera feeds to figure out how to drive safely on the road. The more data AI has access to, the more effectively it can learn a new task and apply that learning to a new environment.

Data velocity refers to the speed of data flow. High rates of data flow are required for real-time applications involving analytics done by AI. For instance, in financial trading, AI algorithms track data streams in real time and make millisecond decisions to trade accordingly.

Data variety refers to the different types of data being generated, collected, and/or stored. Machines that employ AI capabilities can typically work with different types of data – structured data (such as the individual data fields that make up a database), unstructured data (such as narrative text and images), and semi-structured data **JavaScript Object Notation** (**JSON**) files and similar formats that lack rigid constraints on data fields). The range of data types an AI machine can process and learn from has grown along with its capabilities.

For example, NLP capabilities have been extended through our ability to train machines on human-sounding text, and computer vision capabilities have been enhanced by ingesting and learning from images. Within a single organization, **machine learning** (**ML**) systems that deal with customer service messages can now digest text from chat logs, extract voice data from voice calls, and even sense and analyze the sentiment of those messages.

Each of these three data dimensions constantly expands with the rise in AI technologies and every innovation and new business requirement.

There is a reciprocal relationship between data and AI, as advances in the technologies involved with AI dictate how much more data is needed, in terms of volume, velocity, and variety. AI systems are data hungry; algorithms powering AI systems train and learn from very large sets of data and require new, constantly updated data to enhance their enhanced image, speech, and text recognition and make useful predictions.

But much of it is also driven by business innovation; enterprises constantly seek new uses for AI to enhance their competitive positioning, whether to personalize and optimize marketing efforts based on consumer behavior analysis or to improve and automate a manufacturing process with the help of predictive maintenance via AI. This requires richer and higher-quality data.

This rate of business innovation and the widespread and rapid adoption of AI technologies has boosted the increase in data volume, velocity, and variety, where AI systems rely on large, fast-moving, diverse datasets. The more companies innovate with AI solutions and scale their use, the more data these systems produce (and demand), in turn expanding the existing data environment. The intimate connection between AI and data means that strategies to handle the growing volume, velocity, and variety of data will be crucial to meeting the needs of AI applications.

Now that you understand data and its significance and relationships with AI and innovation, you can read about data categories in the next section.

Data Categories

Data types and categories are important for understanding data in this data-driven world. In your professional life, you deal with information in the form of data in systems and applications. You also deal with information in your daily life – from real-time social media platforms and streaming music and movies to playing games and online banking.

All this data makes a lot of noise, so whether you are a business professional, a student, or a curious individual, knowing how data is categorized can help you better navigate the digital landscape.

Data can be categorized as follows:

- **Structured data**
- **Semi-structured data**
- **Unstructured data**

In the following sections, each data categorization will be broken down with simple examples and analogies to everyday use cases you may have encountered, making it easy for those at a base level of understanding to grasp these fundamental concepts.

Structured Data

This data is organized according to some fixed format or predefined structure and stored in a way that can be easily accessed, interpreted, and used.

When data has rows and columns and is arranged in a predefined format with a fixed schema, the data is said to be structured.

It is easier to describe with an example of structured data. A database is one example, but for something more relatable away from your professional life, consider a box of recipe cards.

This could include a card for creating your own pizzas; the card follows a defined specific format. In this example, it could be a format of columns for `Pizza Name`, `Base`, `Sauce`, `Toppings`, and `Vegetarian`. This is illustrated in *Table 1.1*:

Pizza Name	Base	Sauce	Toppings	Vegetarian
Steve's Meat Feast	Stuffed Crust	Tomato	Mozzarella, Pepperoni, Sausage, Ham, Beef	No
Miles Spicier	Deep Pan	Tomato	Mozzarella, Spicy Beef, Peppers, Chili's	No
Better Bianca	Thin and Crispy	Bechamel	Mozzarella, Mushrooms, Broccoli	Yes

Table 1.1: Recipe card analogy to represent structured data

Each card has a recipe (row) in the same structured format, with fields (columns) representing information specific to the pizza recipe. This uniform structure makes it easy to find and follow any recipe; you read all the information contained in the row, but equally, it is also possible to make a decision and selection based on reading information in the column; that is, you want to know which recipes are a vegetarian option or which ones have a deep pan base.

In your professional life, this structured data would be expressed in tabular formats such as spreadsheets (or databases). A spreadsheet example is shown in *Figure 1.2*:

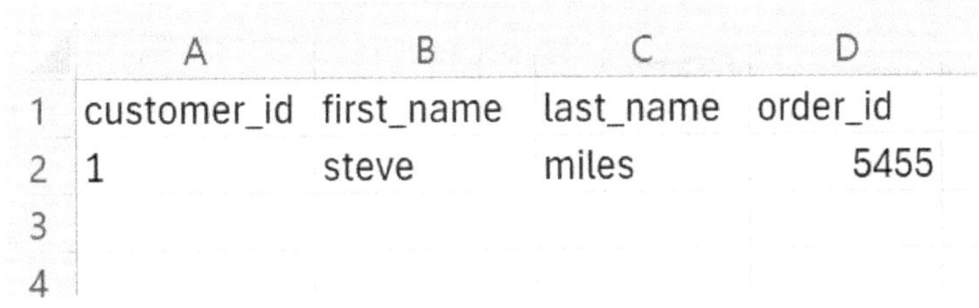

	A	B	C	D
1	customer_id	first_name	last_name	order_id
2	1	steve	miles	5455
3				
4				

Figure 1.2: Structured data representation on a spreadsheet page

The fixed schema components for the structured data shown in *Figure 1.2* represented in a spreadsheet could be defined by tables with a fixed set of defined columns and rows.

The columns are predefined in the fixed schema and have an attribute value for a column corresponding to **text**, **numeric**, **date**, or **Boolean** entry.

An entity (or record) is represented by a row, with the attributes for the entity held in the columns.

Figure 1.3 shows the same structured data entity in *Figure 1.2* represented as a database table:

customer			
customer_id	first_name	last_name	order_id
1	steve	miles	5455

Figure 1.3: Structured data representation in a database table

Once data is placed in tables with attribute headers for each column, database objects are known as keys and can link any table. Data can be retrieved from across tables with **Structured Query Language** (**SQL**) statements; in brief, SQL is a standardized programming language used for managing and manipulating relational databases.

In summary, structured data is like an organized system where every piece of information follows a specific format, making it easy to store, find, and use. Whether it's a spreadsheet, a contact list on your phone, or a box of recipe cards, the key idea is that the information is arranged predefined, consistent, predictable, and simple to find.

Unstructured Data

Unstructured data refers to broad, unspecific data that does not use a specific data model or follow a specific data structure and has no fixed schema. Diverse formats are probably the most distinguishing feature of unstructured data; examples are files, videos, audio, emails, and social media posts. This is stored as binary information, as represented in *Figure 1.4*:

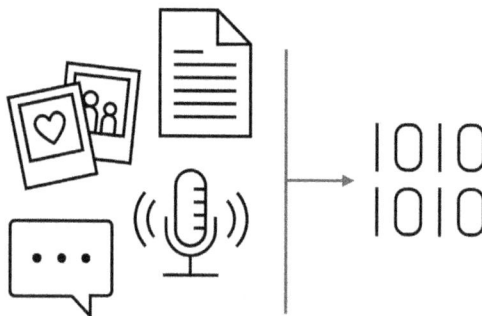

Figure 1.4: Representation of unstructured data

Despite its messy nature, unstructured data can be an incredibly rich source of insights – often in a way that highly structured data cannot deliver.

Here is an analogy to better understand this concept of unstructured data.

Unstructured data is like a messy drawer where things are thrown in without specific organization. It does not follow a predefined format, making it harder to search and analyze than structured data. To continue the analogy, in this messy drawer, you cannot easily find items or know whether something you are looking for is in the drawer. To begin with, when you need something, you must rummage through it because objects are not categorized or labeled.

Moving from the analogy back to the actual topic of data, unstructured data being open to so much diversity definitely allows rich and valuable information to be captured. But at the same time, this makes any standardization quite difficult; unstructured data does not use or even fit into a neat schema, that is, rows and columns of structured data stored in a database. The big advantage of the lack of constraints on unstructured data is that it can capture such rich and deep data.

One of the key challenges of unstructured data is its storage and management. Unstructured data is not well suited for relational databases and thus requires other systems, such as data lakes or NoSQL databases, to accommodate varied data formats. The systems must also be able to scale effectively as the amount of unstructured data is growing quickly due to growing content creation through innovations such as AI and platforms such as social media.

Most unstructured data comprise text-heavy content, such as written documents, emails, and social media posts. This text-based content is rich in context and nuance but requires advanced processing techniques such as **natural language processing (NLP)** to extract meaningful insights.

Since unstructured data files such as video and high-resolution images are large, they require high-density storage facilities and compression mechanisms to compress data files effectively to save space and retrieve them quickly.

Understanding these features highlights unstructured data's challenges and opportunities, emphasizing its importance in modern data analysis and decision-making processes and as a rich source of deep insights and innovation.

In conclusion, unstructured data is a collection of miscellaneous items without a specific order or format. It includes email messages, social media posts, photos, and documents, where the content varies widely and doesn't fit into a neat, predefined structure. This makes unstructured data more challenging to manage and analyze than structured data, which is neatly organized, such as a spreadsheet or a recipe card box.

Semi-structured Data

This data category has a combination of structure and flexibility. Semi-structured data bridges the gap between being structured (where each line of data needs to fit into a specific format, such as the predefined columns of a spreadsheet) and unstructured (where every piece of information is free of any constraints).

An analogy can be used here to explain this concept.

Think of the drawers in a filing cabinet in an office, with folders with different categories marked on each one, such as `Projects`, `Clients`, or `Invoices`.

This filing cabinet represents semi-structured data. Although categorized, it is more flexible and less strict than structured data. Within each folder, any documents or papers in differing formats can be stored; there is no control or definition (schema) on what item can be stored in a folder or information content, but all items in the folder share the common folder label category that provides basic organization to the folder and filing cabinet system.

This represents semi-structured data, which has some organization but also a lot of flexibility. Unlike structured data, it does not fit perfectly into rows and columns. Examples include emails with tags, JSON and **Extensible Markup Language** (**XML**) files, and certain log files. Semi-structured data balances order and flexibility, making it adaptable to various needs.

Semi-structured data uses the base format of structured data but with some flexibility of unstructured data.

A flexible schema is used for properties of variable amounts, that is, not fixed; the properties are represented as key-value pairs.

This makes it more agile to control the organization of information (data) that needs to be represented. It is less complicated than structured data, which has a fixed schema, non-variable column properties (attributes), and an inflexible structure.

The following are examples of semi-structured data:

- **XML files**
- **JSON files**
- **NoSQL databases**
- **Graph databases**

An example is shown in *Figure 1.5* of semi-structured data:

```
1   {
2         "customer_id": "1",
3         "first_name": "steve",
4         "last_name": "miles",
5         "order_id": "5455",
6   }
7
```

Figure 1.5: Semi-structured data representation

The *Figure 1.5* example of semi-structured data is a JSON file with key-value pairs. The advantage is that the data represented is also in a format that can easily be read and interpreted by a human and parsed by a machine.

In summary, semi-structured data exhibits properties of both structured and unstructured data. It generally has a loosely structured schema (such as the folders in a digital photo album or online document storage platform, tags in a typical email client, or folders in a filing cabinet). It is more flexible than a strictly structured data type, and unlike unstructured data, it still provides some level of organization and control.

By understanding these three categories – structured, semi-structured, and unstructured – you can better appreciate the complexities of data management. Each data type has unique characteristics and uses, and knowing how to handle them is essential in our increasingly data-centric world.

Now that you have learned how to describe and categorize data, you will learn about the data storage options available in Azure in the next section.

Identify Options for Data Storage

This section aims to cover the options of how data (information values) can be stored in files and databases. First, you will explore data files and some common formats as part of the exam objectives.

Data Files

Data is organized and stored in a storage solution determined by the data structures discussed in the previous sections. Being aware of the characteristics of the available data storage formats is a necessary skill because each has its benefits and disadvantages, and it is required to meet differing needs and requirements.

In choosing an appropriate data storage solution, you must consider which type of data you have and the application and user requirements, among other things.

Making the right decision enables you to improve data performance based on the key metrics of speed, throughput, latency, scalability, consistency, and reliability.

The following are examples of real-world applications for data performance:

- **E-commerce**: Businesses such as Amazon need high-speed data operations to process millions of transactions, search results for users, and updates on the availability of items in real time.

- **Finance**: Stock-trading platforms must have low latency and high throughput to process real-time transactions and updates and provide real-time trading information.

- **Social media**: Users of Facebook, Twitter, Instagram, and other social media sites expect fast performance; otherwise, they may get frustrated and abandon their visit. These sites' data performance needs to be reasonable and efficient to enable quick creation, sharing, and viewing of content, such as photos, videos, and comments, along with friend and group communication.

- **Healthcare**: Healthcare records depend on high-performance databases to maintain patient records, conduct live monitoring, and guarantee the delivery of critical information on the spot.

In summary, data performance is a crucial aspect of database management that affects how effectively data systems can handle various operations.

The following strategies can be used to efficiently store data, which means organizing data in a way that maximizes storage space while enabling fast access and retrieval:

- **Data compression**: Compressing data reduces the amount of storage space required. There are two types of compression:

  - **Lossless compression**: Reduces the data as much as possible while retaining all the information to reconstruct the original data exactly. Examples are ZIP, GZIP, and PNG (for images).

  - **Lossy compression**: Reduces the size of the data by deleting some information. This might be fine for some types of content, such as images and audio, but not for others, including executable code. JPEG, MP3, and MP4 are examples of this approach.

- **Data storage tiering**: Involves using different types of storage media for different types of data based on access frequency and performance requirements. Frequently accessed data is on faster, more expensive media, for example, storage systems backed by SSDs, while infrequently accessed data, such as archive data, is stored on slower, cheaper media storage systems backed by HDDs.

- **Data normalization**: Normalization in relational databases refers to the process of organizing data to reduce redundancy and dependence. The main strategy is to decompose large tables into smaller tables with relationships to improve data integrity and reduce redundancy.

- **Data indexing**: Indexes speed up data retrieval operations by providing quick access paths to data. However, excessive or poorly designed indexes can consume additional storage space and slow down write operations.

- **Partitioning and sharding**: This involves dividing a database into smaller, more manageable pieces that can be stored across different storage devices. The performance can be improved by reducing the amount of data scanned during queries.

- **Use of appropriate data types**: The right data types (such as integer, character, date and time, Boolean, binary, and UUID) for storing information can save space and improve performance.

- **Schema design**: A well-designed schema reduces redundancy and optimizes data access paths; this includes appropriate primary and foreign keys and avoids nested queries and overly complex joins.

- **Database management practices**: Regular database maintenance, such as regular backups and obsolete data clean-up, can improve storage efficiency.

The following is an example of how data efficiency can be applied.

As a real-world example, **Facebook** uses data sharding to distribute the load to its user base and associated data. User data is equally partitioned and distributed across multiple systems to provide quick responses and uniform load functionality. This enhances performance and scalability, allowing Facebook to cater to enormous data demand. Millions of concurrent users sign in to the system and interact without any issues.

This example demonstrates how effective data storage strategies are implemented in various industries to handle large datasets, ensure quick data retrieval, and maintain cost-efficiency. By adopting techniques such as data compression, sharding, and scalable storage solutions, these companies can meet the demands of their extensive user bases and complex data management needs and ensure that their data systems perform efficiently.

Files can be stored in a number of formats; the following are the most common file formats for these files, which you will learn about in more detail in the next sections:

- **Delimited text**
- **JSON**
- **XML**
- **Binary Large Object (BLOB)**
- **Optimized formats**, such as **Avro**, **ORC**, and **Parquet**

You will now explore these common formats in more detail to prepare for exam questions on them.

Delimited Text

Delimited text is a file format used for storing and exchanging data in which individual data values are separated by a character, known as a **delimiter**. These files are very popular because many software applications and programming languages easily support them. Delimited text files typically use commas, tabs, semicolons, or spaces as delimiters, but most often, the **comma-separated values** (**CSV**) format is used.

Figure 1.6 shows an example of delimited text as a CSV file:

```
1  customer_id,first_name,last_name,order_id,
2  1,steve,miles,5455,
3
4
```

Figure 1.6: CSV file data representation

The delimited text format can be applied in the following ways:

- **Data exchange**: This is one of the most important uses for delimited text files to exchange data between systems, software applications, and platforms; due to their universality, these files can be easily imported and exported.

- **Data storage**: Simple datasets are often stored as delimited text files, frequently for use with spreadsheet software such as Microsoft Excel or Google Sheets.

- **File format**: Delimited data is written one record at a time, following a line-by-line mode. Each line of the file holds one record, separated by one or more delimiters (commas, semicolons, pipes, or tabs). Delimited data is a common choice because you can open these files easily in many data-processing tools or programming languages (such as Python, R, or SQL), which often natively support a read/write operation for such a simple file structure.

- **Configuration file**: Delimited text stores configuration files with a readable structure where settings are stored.

One of the main benefits of delimited text is its generic simplicity; the format is easy to understand and requires no special software to view or modify it. This inherent simplicity, combined with delimited text being simply normal text, further increases its portability.

Delimited text files can be used in a wide variety of software applications on platforms using different operating systems, thus allowing you to transfer them between applications without worrying about compatibility issues. That's why delimited text is one of the most popular file formats for sharing and exchanging data and why every software application employs it, from all sorts of databases to spreadsheets to data analysis applications.

Another important benefit of delimited text is the human readability of the encoded data. This makes it possible to inspect, examine, and troubleshoot data in delimited text using nothing but a text editor.

Delimited text has many advantages, but there are drawbacks, too. One of the most significant drawbacks is that some data types are not supported. All data is stored as text, so values such as dates or numbers always have to be converted from text to the appropriate data type, and this conversion can be performed in code, which can be tedious and troublesome. The inherent limitations of the delimited file's structure, such as a lack of support for complex hierarchical data structures that XML or JSON can express, can also be a hindrance.

In closing, delimited text is an effective and popular way to store and exchange data. The simplicity, readability, and broad support for delimited text make it a great choice for many uses, particularly with simple datasets or when dealing with data that will be used with multiple systems and input devices. A basic understanding of delimited text and how to work with it is important for data management and processing.

JSON

JSON provides a lightweight data interchange format. Its syntax is easy for humans to read and write and for machines to parse and generate. It is language-independent yet also uses conventions familiar to programmers of JavaScript, Perl, Python, and many other languages.

JSON has generally become the preferred format for transferring data between different programming languages and systems. *Figure 1.7* shows an example of a JSON file:

```
1   {
2        "customer_id": "1",
3        "first_name": "steve",
4        "last_name": "miles",
5        "order_id": "5455",
6   }
7
```

Figure 1.7: JSON file data representation

A JSON file has a hierarchical document schema consisting of a collection of objects (data entities) and their attributes stored as key-value pairs.

The keys will always be strings; the values can be strings, numbers, objects, arrays, or null. The data format (or syntax) uses the double quotes symbol, that is, " ", to enclose the keys. The key values are separated by a colon symbol, that is, :. To separate key-value pairs, a comma symbol, that is, , , is used.

Even with minimal documentation or special tools, JSON is easy for non-experts to inspect and understand at a glance, making for a flexible and useful storage format. Readability is especially valuable during the development and debugging stages when a human-readable format can help with spotting errored or incorrect data.

The lightweight characteristic of JSON also makes it suitable for mobile and embedded applications, where bandwidth and storage might be limited. By keeping data sizes small, JSON helps ensure faster load times and a better user experience.

Finally, JSON's interoperability is a major benefit, particularly in an era where systems integration and communication are paramount. The wide acceptance and standardization of JSON ensure that it is recognized and utilized by a vast array of systems, platforms, and technologies.

XML

XML is an open standard for representing documents and sharing information in a human-readable, text-based, and machine-readable format.

It is used as a general-purpose markup language when data exchange from one application to another must be flexible – where it is not known within the application what data is being exchanged – or where a more flexible format than the original one is needed for further data manipulation.

It aims to simplify the sharing of both the format and the data on the web, on intranets, and elsewhere. Unlike HTML, which is used for display purposes, XML is used for data coding, storage, and transportation.

Markup tags and attributes represent the structure and content of an XML data file; these make the information accessible to parse and manipulate.

An XML document can be viewed as a **tree structure**, with a single **root element** containing all other elements.

XML's document structure corresponds to the structure of a tree, with a unique root element containing all other elements.

The elements are enclosed within `<element>` and `</element>` tags and are hierarchical; an element can contain other elements, attributes, or text.

Figure 1.8 shows an example of an XML file structure:

```
1  <root>
2      <customer_id>1</customer_id>
3      <first_name>steve</first_name>
4      <last_name>miles</last_name>
5      <order_id>5455</order_id>
6  </root>
7
```

Figure 1.8: XML file data representation

The key **differentiator** between JSON and XML when looking at a block of code text to determine the file data representation is that JSON uses double-quote, " ", symbols to enclose the keys, whereas XML uses **markup tags**, < > .JSON does not use tags, which makes it more compact and easier to read for humans. JSON can represent the same data in smaller files for faster transfer.

XML has several key features that contribute to its widespread use and versatility. The key is that XML is extensible, so you can define custom tags and use them to represent complex data structures. This flexibility makes XML suitable for many different tasks and applications. XML documents are self-describing.

Another notable feature is XML's hierarchical structure, which organizes data in a tree-like format. XML's platform independence is a critical advantage, as it ensures that XML documents can be used across different systems and technologies without compatibility issues. This makes XML an ideal format for data interchange in diverse environments.

In summary, XML is a powerful and flexible format for encoding documents and data. Its extensibility, hierarchical structure, and platform independence make it a versatile tool for data storage, transport, and interchange. XML's standardized syntax and human readability further contribute to its widespread adoption across various domains, from web development to data management.

Binary Large Object (BLOB)

A **BLOB** unstructured data file format stores large amounts of binary data, including multimedia files such as images, audio, video, and other types of large binary files. Unlike text-based formats, BLOBs can store data in a raw, unstructured form, making them ideal for handling any file that needs to be stored in a database or filesystem without regard to its internal structure.

Figure 1.9 shows an example of binary data as a **BLOB** file:

```
1  00100000 00100000 00111100 01100011 01110101 01110011 01110100 01101111
   01101101 01100101 01110010 01011111 01101001 01100100 00111110 00110001
   00111100 00101111 01100011 01110101 01110011 01110100 01101111 01101101
   01100101 01110010 01011111 01101001 01100100 00111110 00001010 00100000
   00100000 00100000 00100000 00111100 01100110 01101001 01101110 01110011
   01110100 01011111 01101110 01100000 01101101 01100101 00111110 01110011
   01110100 01100001 01110110 01100101 00111100 00101111 01100110 01101001
   01110010 01110011 01110100 01011111 01101110 01100001 01101101 01100101
   00111110 00001010 00100000 00100000 00100000 00100000 00111100 01101100
   01100001 01110011 01110100 01011111 01101110 01100001 01101101 01100101
   00111110 01101101 01101001 01101100 01100101 01110011 00111100 00101111
   01101100 01100001 01110011 01110100 01011111 01101110 01100001 01101101
   01100101 00111110 00001010 00100000 00100000 00100000 00100000 00111100
   01101111 01110010 01100100 01100101 01110010 01011111 01101001 01100100
   00111110 00110101 00110100 00110101 00110101 00111100 00101111 01101111
   01110010 01100100 01100101 01110010 01011111 01101001 01100100 00111110
2
```

Figure 1.9: BLOB file data representation

As shown in *Figure 1.9*, BLOB files, being binary data, have no human-readable data representation.

Some key features of the BLOB file format make it versatile and usable for many applications. The most important one is that a BLOB can be used for any binary file. This means that it can be used to store images, audio, and video files. However, it can also be used for any other type of binary file, including any file the application needs to manage.

Because the extremely flexible nature of BLOB makes it well suited to the storage of large, binary datasets, it is used across a broad cross-section of industries and applications. A notably widespread use of BLOBs is in multimedia store-and-retrieve systems, which use BLOBs to store, retrieve, and manage datasets that include images, audio files, videos, and other media resources.

In addition, BLOBs serve as valuable backup and archiving tools, enabling organizations to store backups of databases, filesystems, and application data. By storing backups as BLOBs, organizations can ensure data integrity, accessibility, and compliance with regulatory requirements.

Overall, BLOB's versatility, scalability, and efficiency make it indispensable for a wide range of applications, including multimedia storage, document management, scientific research, backup and archiving, and software distribution. As organizations continue to generate and manage increasingly large volumes of binary data, BLOBs will remain essential tools for effectively handling and storing such data.

Optimized: Avro, ORC, and Parquet

Optimized file formats such as **Avro**, **Optimized Row Columnar** (**ORC**)**,** and **Parquet** are specifically designed for the efficient storage and processing of large datasets, particularly in the context of big data analytics and distributed computing frameworks. These formats offer various optimizations, such as columnar storage, compression techniques, and schema evolution support, to achieve high performance and scalability.

You can now explore each of them:

- **Avro**: Uses a row file format. This is a **binary serialization** format developed within the **Apache Hadoop** project. It features a compact binary encoding scheme, schema evolution support, and rich data structures such as records, arrays, and maps. Avro's schema evolution capabilities allow data schemas to evolve over time without breaking compatibility, making it suitable for evolving data pipelines and data lakes. Avro is commonly used in **Apache Kafka**, **Apache Hadoop**, and other big data frameworks for data serialization, messaging, and storage.

  This file format is designed to be compact and efficient for storing and transmitting large datasets, making it suitable for distributed computing environments. It stores data as binary information in rows and uses a self-describing JSON-based schema that defines the data structure.

- **ORC**: Uses a column file format. This columnar storage layout improves query performance by minimizing I/O operations and reducing data transfer overhead. ORC also supports advanced features such as predicate pushdown and vectorized query execution, making it well-suited to analytical workloads in data warehouses and data lakes. This is a highly efficient file format used for big datasets. The aim is to optimize performance and reduce storage requirements by compressing data and storing it as columns rather than rows, allowing faster queries and analysis. ORC files use file stripes, which are units of data storage. Each stripe contains a portion of the data for all columns and is used to organize data efficiently for parallel processing and I/O operations.

- **Parquet**: Uses a column file format. This columnar storage file format emerged from the Apache Hadoop ecosystem. Like ORC, it stores data in per-column layouts that promote efficient compression and per-column encoding schemes. Parquet files are highly compressed and splittable, allowing them to be broken up into small chunks that can be processed in parallel within a distributed computing cluster, such as an Apache Spark or Apache Hive environment. This file format is used to store large amounts of datasets. This format provides efficient performance and storage optimization, allowing related data to be compressed and compact. Parquet files are often used to process massive datasets with distributed computing. This format has increased in popularity and is becoming widely used in the data analytics community for data warehousing and data lake architectures because it can quickly store and process structured and unstructured data.

An example of where this file format is used is in **OneLake**, part of **Microsoft Fabric**, a SaaS-based unified data analytics platform.

In summary, Avro, ORC, and Parquet are optimized file formats designed to meet the performance and scalability requirements of big data analytics and distributed computing applications. Each format offers unique features and optimizations tailored to specific use cases, making them valuable tools for storing, processing, and analyzing large datasets in modern data architectures.

In this section on data file formats, you read about each file format for storing data, including delimited text, JSON, BLOB, and optimized formats such as Avro, ORC, and Parquet. To close this topic on data stores, you will look at what a database is and the different types.

Databases

A database is an organized collection of stored electronic data that can be accessed, searched for, and used to provide information.

The primary functions of a database are to store and retrieve data, making sure that the data is valid and security features are in place while being scalable for growing amounts of data and evolving demand by people wanting to use that data and similar systems, allowing for concurrent access by different users. Essentially, this is a massive and pervasive application.

Databases can be of the following two types:

- **Relational**: Uses a fixed schema that stores structured data
- **Non-relational**: Uses a flexible schema or no schema that stores semi-structured or unstructured data; provides self-describing entities

These database types are essential in banking, business, education, e-commerce, finances, government functions, and healthcare.

The three core components of a database are the **data**, the **data structure**, which consists of a schema, tables, indexes, and views, and the **operational components**, which are the **Database Management System (DBMS)**, the data queries, and data transactions (data reads and writes).

The data is everything that can be stored: text, numbers, images, and everything in between. When you look at the database structure aspects, the schema describes the structure of tables, fields, relationships, and constraints that underlie the storage and manipulation of data in a database. The fundamental building block of a database is a table, in which an entity (such as a customer, an order, or a product) is described as a set of rows and columns. Indexes speed up data retrieval but take up more storage and must be maintained. At the same time, views are virtual tables created by asking for data distributed across one or more tables, which allows for simpler queries that otherwise may be quite complex, and from a security aspect, each view can limit access to specific table data.

You interface with a DBMS (software that provides tools to store, manage, and retrieve data in databases efficiently) to manage or perform operations on data, such as adding new records, querying for certain data, modifying data, or deleting data. Common examples of database DBMS software include MySQL, PostgreSQL, Oracle, and Microsoft SQL Server.

With queries, users ask questions, typically written in SQL, that require the database to retrieve the data they are interested in, and transactions ensure that data integrity is maintained by treating sequences of operations as atomic units that behave in accordance with the **Atomicity, Consistency, Isolation, Durability (ACID)** properties. For clarity, data transactions are sequences of operations performed on a database that are treated as a single, indivisible unit to ensure data integrity and consistency.

In business operations, such as CRM, supply chain management, **Enterprise Resource Planning (ERP)**, and even e-commerce, databases are used to support product inventories, customer information, orders, transactions, and much more. Healthcare heavily relies on databases for health record and patient management, including vital patient information, medical history, and treatment plans. Financial institutions use them to process transactions, manage customer accounts, and look out for fraud. Finally, educational institutions use databases to manage student records, course enrollments, and academic performance.

Databases are fundamental to modern information systems, offering structured storage, efficient data retrieval, and robust data management. Their applications are vast and varied, underscoring their critical role in supporting essential operations and enhancing decision-making across multiple industries.

You will now explore the types of databases in the following sections.

Relational Databases

A relational database provides a structured and organized way to store, find, and connect different pieces of information, making it a powerful tool for managing data efficiently.

A relational database comprises tables of rows and columns that organize the data. A record is represented by a row; a specific attribute of that record is represented by a column.

One of the great strengths of relational databases is that you can link information across tables; that is, you can have a database table called `Customers` and one called `Orders`, and you can link the two to determine which customer made which order.

You often say you are "querying" the database when you ask it a question. For instance, you could query the database and say, "Show me all the people who bought something this month." The database searches its tables and columns for the answer; this displays (returns) the rows that match the criteria.

In essence, this type of database is like an organized digital filing system where you can store, find, and connect all kinds of information.

Some key characteristics of relational databases are as follows:

- **Structured data model**: This uses a fixed schema; that is, it specifies the tables and their rows, columns, and table relationships that can be used before any data can be written to the database tables. This is called schema on write.

- **Tables and relationships**: Tables with distinct names are created, such as `Customer`, `Products`, `Orders`, and `Employees`, that identify the type of information that can be found in that table; think of these like "headings" in a book, "labels" on a box file, or the name you may give a spreadsheet page.

- **Keys**: These are database table column objects that establish relationships between tables. Through **primary keys** and **foreign keys**, they specify the relationship between data in one table column and data in another.

- **Normalization**: To reduce data values, **duplication data** is normalized; this means that the attributes and different "dimensions" of the data are split into individual tables that are then linked to maintaining their relationship. The intention is the opposite of aggregating data.

- **SQL**: This common query language is used to interact with the data; you can use this to enter, manipulate, and retrieve data stored in the tables.

- **ACID properties**: Relational databases conform to the ACID properties. These properties will be explained in more detail later in the section.

- **Data integrity and constraints**: The data integrity in relational databases is maintained by different constraints, such as primary key constraints, foreign key constraints, unique constraints, and check constraints. These ensure the data is correct (integrity) and consistent (conformity).

- **Indexing and query optimization**: Relational databases have high transactional write operations and often use indexes for faster data retrieval, especially for big datasets. Indexes help obtain effective access routes to data via certain conditions, enhancing query retrieval.

The relationship between tables within a relational database is illustrated in *Figure 1.10*:

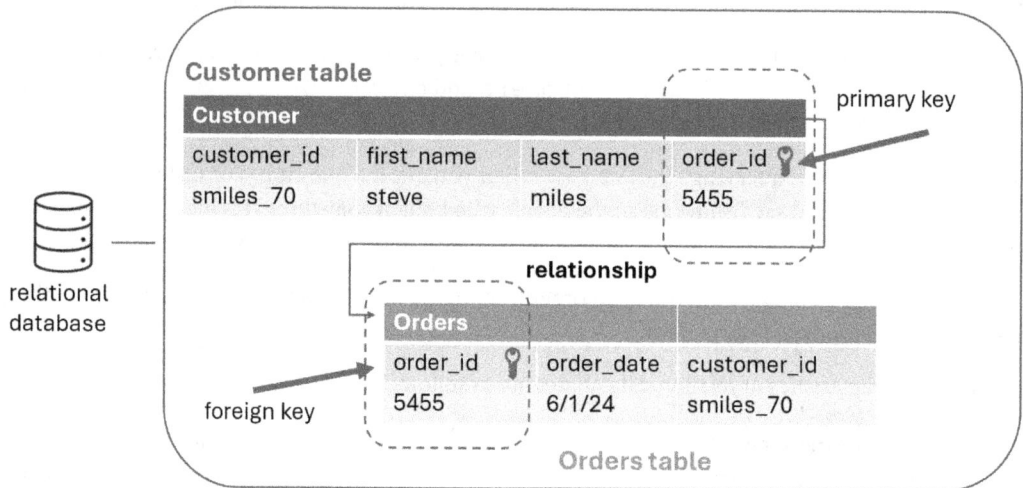

Figure 1.10: Relational tables in a database with keys

The popularity of relational databases is because of their flexibility and reliability and the fact that they can handle intricate data relationships in several fields and applications. Some popular RDBM solutions include Microsoft SQL, MySQL, and PostgreSQL.

Non-relational Databases

A non-relational or NoSQL database operates differently from a relational database in structure and organization. Unlike relational databases that use tables with predefined schemas, non-relational databases employ a more flexible approach, allowing for storing and retrieving data in various formats without strict adherence to a fixed schema.

For non-relational databases, that data might be stored in collections (in the case of document databases) or documents (in the case of graph databases). What's stored in these collections and documents can be of various data types (key-value pairs, JSON documents, or graph structures), making non-relational databases easier to use when the data can vary over time.

Some key characteristics of non-relational databases are as follows:

- **Schema-less** or **flexible schema**: There is no fixed schema on the data. This type of structure allows every record or document in the database to have its structure and fields added or removed without altering a **predefined schema**, referred to as self-describing entities. (Unlike relational databases that impose the fixed schema defined by tables, columns, and relationships, non-relational databases typically support flexible or schema-less data models.)

- **Horizontal scalability**: Non-relational databases are implemented in horizontal scaling; in other words, they can accommodate large amounts of data and high traffic by adding more commodity servers or nodes in the database cluster. This is opposed to the vertical scaling commonly introduced with relational databases, where a single server is given more resources.

- **Distributed architecture**: A distributed architecture designs non-relational databases where the data is stored in multiple servers or nodes. This architecture improves fault tolerance, resiliency, and performance by sharing data processing and storage works among the cluster.

- **High performance**: Most **NoSQL databases** are designed for particular use cases or data models, including **key-value stores**, **document stores**, **column-family stores**, or **graph databases**. Compared to regular relational databases, this specialization usually relates to high performance when dealing with particular query types and operations.

- **Support for unstructured and semi-structured data**: Unstructured and semi-structured data formats are better processed by non-relational databases, including JSON and XML.

- **Eventual consistency**: Availability and partition tolerance are the major priorities of some NoSQL databases over strict consistency.

- **Scalability patterns**: NoSQL databases typically use scaling patterns, such as **sharding**, **replication**, and **partitioning**, to distribute data and load across the cluster more efficiently. Such patterns enable better utilization of resources and performance enhancement as the database grows.

Several non-relational or NoSQL databases are designed to address specific use cases and data storage needs. Some common types of non-relational databases include the following:

- **Key-value databases**: Data entities are represented and stored as key and value pairs; no schema is applied. They are suited for high-performance use cases such as **caching**, where quick reads and writes are needed, not for complex queries or relationships. An example is **Redis**.

- **Document databases**: A document database is schema-less and stores data as semi-structured documents. These databases store data in flexible, semi-structured document formats such as JSON or XML. An example is **MongoDB**.

- **Column databases**: Also known as columnar databases, a column database stores data entities in columns instead of rows, making them efficient for read-heavy analytical tasks over large datasets. An example is **Apache Cassandra**.

- **Graph databases**: These databases are optimized for managing and querying **graph data** structures. **Nodes** and **edges** (vertices) are used to capture entity relationships. They are ideal for complex data relationships and optimized for traversing relationships between nodes. An example is **Apache Gremlin**.

In summary, non-relational databases are often chosen for their flexibility, scalability, and ability to handle unstructured or semi-structured data more effectively than relational databases. They are commonly used in modern web applications, big data analytics, real-time data processing, and other scenarios where traditional relational databases may not be the most suitable solution.

Now that you have learned about relational and non-relational database characteristics and types of databases, it is time to learn about some common data workloads.

Describe Common Data Workloads

This section will help you understand the data processing systems used in databases. These systems are essential for managing data within organizations and are designed to serve different purposes to support transactional and analytical workloads.

The two distinct types of systems used for processing data are as follows:

- **Online Transactional Processing (OLTP)** data systems for operational data
- **Online Analytical Processing (OLAP)** data systems for analytical data

The following two sections in this chapter look at the features of each of these and the scenarios of when one may be used over the other.

Features of Transactional Workloads

Application workloads use OLTP data processing solutions optimized for storing operational data in a relational database. The applications that process operational business data are often called **line of business (LOB)** applications.

Transactional workloads prioritize data consistency, integrity, and reliability, ensuring that database transactions are executed correctly and that data remains accurate and consistent even in a concurrent and distributed environment.

Key characteristics of OLTP systems are as follows:

- **Schema**: A fixed schema organizes and controls the data written into the table. These predefined schemas must exist before this data can be stored, referred to as **schema on write**.

- **Transactional operations**: Transactions are stored as data in the database tables. The database operations are a mix of medium read and heavy write; that is, retrieving a catalog entry from a `Product` table uses a read operation, while logging a purchase in the `Order` table uses a write operation.

- **Real-time processing**: Data is written instantly with minimal latency.

- **ACID properties**: OLTP systems implement ACID-based transactions, which are properties that guarantee the consistency and integrity of data. Each of these properties is described as follows:

  - **Atomicity**: This property requires that transactions, which are single units of work, are fully completed or fail/rolled back completely.

  - **Consistency**: This property requires that the database state remains valid before and after any transactions occur.

  - **Isolation**: This property requires that transactions operate independently to prevent interference.

  - **Durability**: This property requires that transactions committed are permanently saved to the database, including in system failure states.

- **Normalized data structure**: This is used for effective utilization and conservation of storage space in OLTP systems; this reduces redundancy in data stored and improves performance.

The Azure data services relational databases that can be used for OLTP include Azure SQL, Azure MySQL, Azure Database for MariaDB, and Azure Database for PostgreSQL. You will learn more about these Azure data services in *Chapter 3, Describe Relational Azure Data Services*.

In this section, you learned about the features of transactional workloads; the following section looks at analytical workloads.

Features of Analytical Workloads

Analytical workloads focus on processing and analyzing large volumes of data to uncover insights and trends that drive informed decision-making and strategic planning. Organizations can extract actionable insights from their data by leveraging scalable infrastructure, parallel processing techniques, and advanced analytics tools to gain a competitive edge and drive business growth.

OLAP refers to data processing systems that use data stores optimized for analytical tasks. These data stores can be a **data warehouse**, **data lake**, or **lakehouse**. You will explore each data store in more detail in *Chapter 6, Describe the Common Elements of Large-Scale Analytics*.

Operational data is extracted from various operational data stores (**sources**) and is loaded into one or more analytical data stores (**sinks**) where analytics tasks or ML can be performed on the ingested operational data to output decision-making insights, data visualizations, or models for further consumption. One further aspect is **data transformations**; this ensures that the ingested data is in the right structure and of the right quality for ML tasks, modeling, analysis, and reporting (garbage in = garbage out).

The transformation process includes tasks such as filtering, manipulating, changing, cleansing, and enriching the data to fit requirements or needs.

The critical question is, at what stage of the data flow do the transformations happen? Do the transformations happen before or after the load?

- If the transformations happen "before" the data is ingested into the analytical data store, this is an **Extract, Transform, and Load** (ETL) approach.

- If the transformations happen "after" the data is ingested into the analytical data store, this is an **Extract, Load, and Transform (ELT)** approach.

These transformation tasks are abbreviated to ETL and ELT and are the responsibilities of the data engineer. In the next section, *Identify Roles and Responsibilities for Data Workloads*, you will discover more about this topic.

The Azure data services that perform these tasks include **Azure Data Factory**, **Azure Synapse Analytics**, **Azure Databricks**, and **Azure HDInsight**. You will discover more about these Azure data services in *Chapter 6, Describe the Common Elements of Large-Scale Analytics*.

Key characteristics of OLAP systems are as follows:

- **Schema**: OLAP uses schema on read, which creates the schema as data is read from tables; the organization and structure are not predefined in a schema. OLTP table data is denormalized and aggregated for loading into analytical and ML models.

- **Interactive querying**: OLAP tools give users an interactive interface to query and explore data and perform operations such as slice, dice, roll-up, drill down, pivot, and ad hoc analysis to answer specific business questions.

- **Optimized for read operations**: OLAP systems must be able to perform read operations efficiently.

- **Aggregation**: OLAP is used in data warehousing technologies as a sink for highly aggregated large volumes of structural and historical data optimized for data analysis and read operations.

This section on OLTP systems for processing analytical workloads concludes the *Describe Common Data Workloads* exam skills measured area. Next, you will move on to a different skill area of the exam that covers data workloads, roles, and responsibilities.

Identify Roles and Responsibilities for Data Workloads

Within organizations, various roles are related to managing and leveraging data effectively. These data roles within organizations use many tools and models to draw meaningful insights that help organizations make informed decisions in their respective domains.

As part of the *Identify Roles and Responsibilities for Data Workloads* skills measured area for the DP-900 Azure Data Fundamentals exam, you will need to understand the following key data job roles:

- **Database administrators**
- **Data engineers**
- **Data analysts**

The following sections look at each of these key data job roles.

Responsibilities of Database Administrators

Database administrators are responsible for the management, security, and optimization of databases in an organization to guarantee the reliability and performance of data systems.

The responsibilities of the role of a database administrator can include the following:

- Creating, configuring, and managing transactional databases
- Security, identity, and access management
- Availability, backup, and disaster recovery
- Monitoring and optimization
- Supporting and troubleshooting transactional databases

Database administrators will perform tasks with relational databases such as Microsoft SQL and open source databases such as MySQL, MariaDB, and PostgreSQL. You will discover more about these Azure data services in *Chapter 3, Describe Relational Azure Data Services*.

Next, you will look at the responsibilities of the data engineer role.

Responsibilities of Data Engineers

Data engineers create and maintain data infrastructure and pipelines to ensure an organization can make data-informed decisions and perform data analysis.

The responsibilities of the role of a data engineer can include the following:

- Creating analytical workload data stores
- ETL/ELT processes including:
  - Data pipelines
  - Data integrations
  - Data cleansing and transformations

Data engineers will perform tasks using Azure Data Factory, Azure Synapse Analytics, Azure HDInsight, Azure Databricks, Azure Stream Analytics, Microsoft Purview, and Microsoft Fabric. You will discover more about these Azure data services in *Chapter 6, Describe the Common Elements of Large-Scale Analytics*.

With the data engineers' responsibilities now understood, you will read about the data analysts' responsibilities.

Responsibilities of Data Analysts

Data analysts utilize data insights to inform business decisions and strategies, giving companies a competitive advantage and helping them achieve their objectives.

The responsibilities of the role of a data analyst can include the following:

- Analytical modeling
- Data exploring
- Data reporting
- Data visualization

Data analysts will perform tasks using Azure Synapse Analytics, Azure Databricks, Azure Data Explorer, and Microsoft Fabric. You will discover more about these Azure data services in *Chapter 6, Describe the Common Elements of Large-Scale Analytics*.

This concludes this section for the *DP-900 Azure Data Fundamentals* exam skills measured area *Identify Roles and Responsibilities for Data Workloads*. This section also concludes the learning content for this chapter. Next, what you have learned in this chapter will be summarized for you.

Summary

This chapter included complete coverage of the *DP-900 Azure Data Fundamentals* exam skills measured area *Describe Core Data Concepts*.

In this chapter, you learned about core data concepts, how data can be represented, data storage options, common data workloads, and how to identify roles and responsibilities for data workloads.

The next chapter will cover relational concepts to follow the skills measured sequence in *Microsoft's study guide for the DP-900 exam*.

Additional Reading

This section provides links to additional exam information and study references:

- *DP-900 - Microsoft Azure Data Fundamentals study guide*: https://learn.microsoft.com/en-us/credentials/certifications/resources/study-guides/dp-900

- *DP-900 - Microsoft Azure Data Fundamentals self-directed learning*: https://learn.microsoft.com/en-us/training/modules/explore-core-data-concepts/

- Choose a data storage approach in Azure: https://learn.microsoft.com/en-us/training/modules/choose-storage-approach-in-azure/

Exam Readiness Drill – Chapter Review Questions

Apart from a solid understanding of key concepts, being able to think quickly under time pressure is a skill that will help you ace your certification exam. That is why working on these skills early on in your learning journey is key.

Chapter review questions are designed to improve your test-taking skills progressively with each chapter you learn and review your understanding of key concepts in the chapter at the same time. You'll find these at the end of each chapter.

How to Access These Materials

To learn how to access these resources, head over to the chapter titled *Chapter 9, Accessing the Online Resources.*

To open the Chapter Review Questions for this chapter, perform the following steps:

1. Click the link – `https://packt.link/DP900Ch01`.

    Alternatively, you can scan the following **QR code** (*Figure 1.11*):

Figure 1.11: QR code that opens Chapter Review Questions for logged-in users

2. Once you log in, you'll see a page similar to the one shown in *Figure 1.12*:

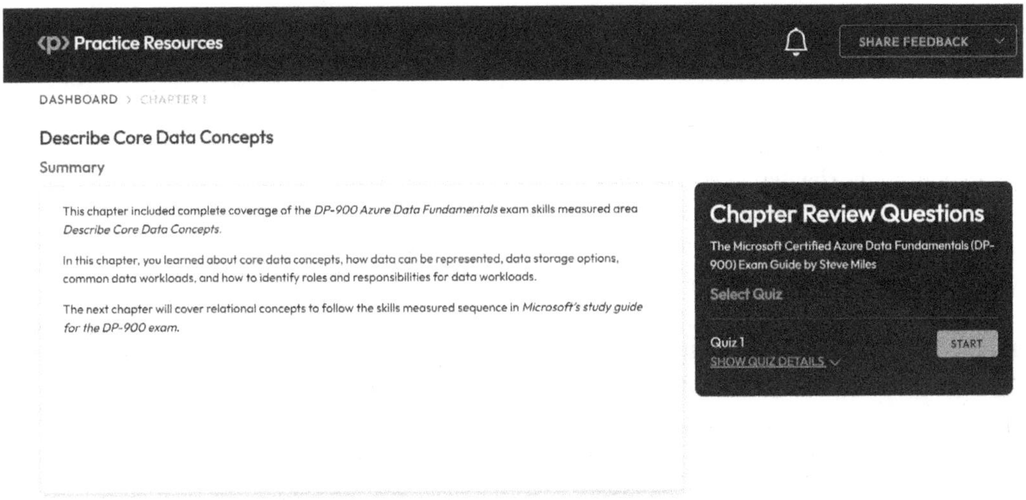

Figure 1.12: Chapter Review Questions for Chapter 1

3. Once ready, start the following practice drills, re-attempting the quiz multiple times.

Exam Readiness Drill

For the first three attempts, don't worry about the time limit.

ATTEMPT 1

The first time, aim for at least **40%**. Look at the answers you got wrong and read the relevant sections in the chapter again to fix your learning gaps.

ATTEMPT 2

The second time, aim for at least **60%**. Look at the answers you got wrong and read the relevant sections in the chapter again to fix any remaining learning gaps.

ATTEMPT 3

The third time, aim for at least **75%**. Once you score 75% or more, you start working on your timing.

> Tip
>
> You may take more than **three** attempts to reach 75%. That's okay. Just review the relevant sections in the chapter till you get there.

Working On Timing

Target: Your aim is to keep the score the same while trying to answer these questions as quickly as possible. Here's an example of how your next attempts should look like:

Attempt	Score	Time Taken
Attempt 5	77%	21 mins 30 seconds
Attempt 6	78%	18 mins 34 seconds
Attempt 7	76%	14 mins 44 seconds

Table 1.2: Sample timing practice drills on the online platform

> Note
>
> The time limits shown in the above table are just examples. Set your own time limits with each attempt based on the time limit of the quiz on the website.

With each new attempt, your score should stay above **75%** while your "time taken" to complete should "decrease". Repeat as many attempts as you want till you feel confident dealing with the time pressure.

2

Describe Relational Concepts

Imagine you were putting together a library. To record the details of thousands of books, authors, and subjects, you would need a highly organized system to find and keep track of everything. Each book must be entered onto the correct shelf for the correct subject, associated with a sub-genre, and stored by some hierarchy, such as having every book by an author grouped or in alphabetical order by title.

Relational data is a lot like this; it is all about organizing and structuring data according to certain rules and principles (a schema) that guarantee consistency, integrity, and efficiency. This type of data is found in various applications and environments, including e-commerce platforms, employee and healthcare records, financial institutions, and so on.

In this chapter, you will delve into the fundamental relational data concepts essential for the DP-900 Azure Data Fundamentals exam. Understanding these concepts is crucial for passing the exam and applying these principles in real-world scenarios where efficient data management is vital.

You will cover several key topics:

- Identify features of relational data
- Describe normalization and why it is used
- Identify common **structured query language (SQL)** statements
- Identify common database objects

The topics are related to DP-900 Azure Data Fundamentals, a certification exam covering skills for recognizing relational data considerations on Azure. This corresponds to about 20–25% of the total measured skills. Completing this chapter ensures you have the skills for identifying relational data on Azure and are ready for the certification exam.

This first part of the chapter is intended to help you get acquainted with the concept of relational data/ database identity and describe these aspects.

You will start by exploring the features of relational data and how these are the foundation for efficient and effective data management systems.

Identify Features of Relational Data

In the first section of this chapter, you will learn what relational data is and the features of relational databases.

It can be helpful to think of the library analogy for relational data, where each book has a unique identification number (e.g., ISBN) and belongs to a particular category (e.g., fiction versus non-fiction). The library tables can be arranged into shelves, each of which would be a table in the system. For each category of books, there will be a table showing all the books. Each shelf will have the books placed into rows and columns in the order in which each book is placed on the shelf. Each row and column on the shelf is associated with exactly one attribute, just like the attributes of a table that come together to relate one instance of the entity (e.g., the book) in the relational database. The library can have a catalog listing all the books and their data, as many tables contain columns listing data associated with specific books. The library can also have data-capturing relationships between books, such as those written by the same author, those in the same series, those about the same topic, in the same category, and so on. Each of these relationships can be seen in the relational database as a reference key embedded in a table that links data between different tables.

Relational data can be defined as data that is organized in such a way as to define the connections and associations between different data points or entities. This type of data structure is commonly used in relational databases to represent the relationships between various pieces of information. It allows for the efficient retrieval and manipulation of data based on these defined relationships and is a fundamental concept in data management and database design.

Some examples of relational data are as follows:

- **Customer information**: Each customer can have an ID number, name, address, phone number, email, and so on. These can be stored in a table called customer, where each row represents a customer, and each column represents an attribute.

- **Employee records**: Each employee can have an ID number, name, department, position, salary, and so on. These can be stored in a table called Employees, where each row represents an employee, and each column represents an attribute.

Data is held in **tables**, which comprise **rows** and **columns**. *Figure 2.1* illustrates a table:

customer			
CustomerID	Title	Surname	Department
001	Mr	Miles	Engineering
002	Mr	Bloggs	Operations
003	Mrs	Miggins	Retail

Figure 2.1 – Table example

The following are the components of tables:

- Each row is an instance of a data entity

- Each data entity can have several attributes assigned; each is defined in a separate column, named with attribute headers

- A datatype is assigned to each column

- Keys are used to uniquely identify a row in a database; they can link tables through a relationship

Figure 2.2 illustrates relational data components:

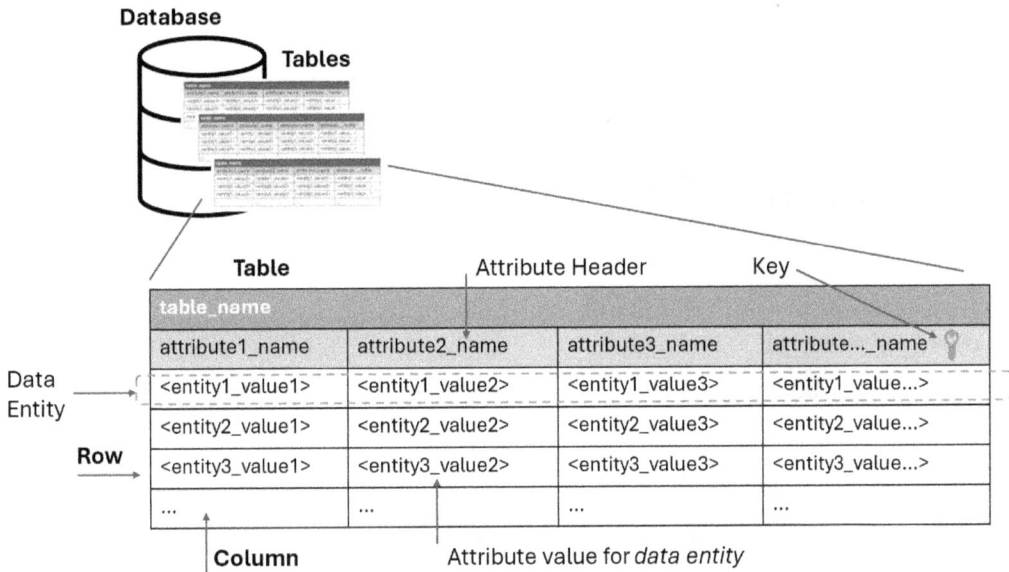

Figure 2.2 – Relational data components

There may be multiple **tables** for customers, orders, products, and so on, each with specific **attributes** for the data **entity**.

The customer table data entities may have attributes that include ID number, name, and orders; for the Orders table, the data entities may have order ID, order date, and customer ID attributes; for the Line Item table, the data entities may have order id, item id, and item qty; and for the product table, ID, name, and price. These tables have relationships that are based on related attributes that have keys that connect tables.

The `Customer` table may use the field for customer ID numbers, which would also appear in the `Orders` table to associate the sales records with the customers. The linking of the tables allows easy access and analysis of data from several tables and gives insight into how different data entities are related. *Figure 2.3* illustrates the points that were covered:

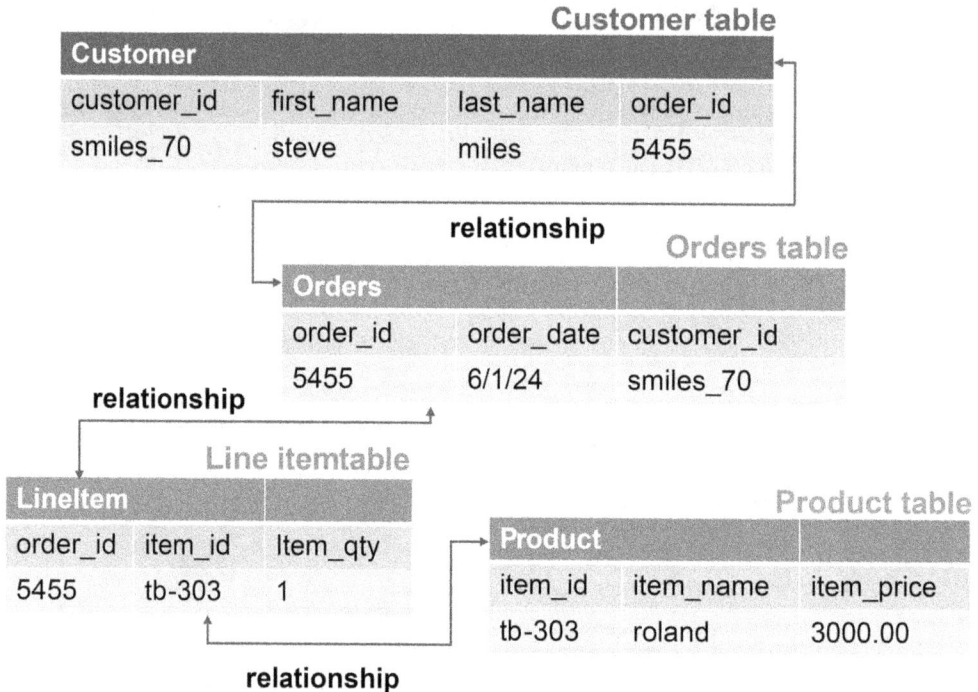

Figure 2.3 – Database table relationships

Relational databases are popular in modern applications and can store different kinds of data, including financial data, customer details, product catalogs, and so on.

Relational databases enable organizations to organize their data consistently and clearly define relationships between different data entities. This helps make modern systems effective and allows data-driven decision-making.

You will now learn about the characteristics of relational databases.

Relational Database Characteristics

A relational database holds the structured relational data in a system accessed by transactional and analytical workloads.

Structured relational data is organized in tables with rows and columns. Each row represents an entity or an instance of data. Relational databases organize information based on structural relationships, so searching for clusters of related information is easy. Structured relational data supports the following two types of workloads:

- **Transactional workload**: The data is frequently added or changed, such as in online transactions, inventory management, or bank operations

- **Analytical workload**: The data is frequently queried or analyzed from large amounts of information, such as in business intelligence, reporting, or data mining

Relational databases have the following characteristics:

- Data entities are stored as structured data

- A fixed schema provides a structure that guarantees strong consistency

- Data entities are stored as data values in tables, a tabular format that consists of rows and columns

- Rows represent data entities; one row per data entity instance

- The same columns are used for each row in the table

- A datatype is assigned to each column in the table; entries can have a value of NULL

- Tables have no duplicate information

- Tables can reference other tables by using keys to provide the relationships

An example of a relational database would be **Microsoft SQL**; this can be implemented as an IaaS or PaaS service on Azure. You will discover the methods for implementing Microsoft SQL for Azure in *Chapter 3, Describe Relational Azure Data Services.*

Now that you understand the features of relational data, you will learn about data normalization.

Describe Normalization and Why It Is Used

Imagine you have a closet full of clothes that are messy, disorganized, and mixed up. It would be hard to find what you need, and you might buy duplicates or lose track of what you have. You would also waste a lot of space, make your closet look cluttered, and spend a lot of time searching for what you are looking for and putting together an outfit.

To solve this problem, you could apply the principles of normalization to your closet. You could sort your clothes by type, such as shirts, trousers, dresses, shoes, ties, scarves, and so on. Then, you could create separate hanging rails, shelves, or drawers for each type of clothing and categorize them accordingly. This way, you would avoid repetition, remove redundancy, and make your closet more efficient, organized, and easy to select for any occasion or need.

Normalization in relational data is like organizing your closet but for data. By moving data into separate tables and columns based on their attributes, you can avoid storing the same information in multiple places, reduce the risk of errors or inconsistencies, and improve the performance and quality of your database. Normalization is used to make data more manageable, reliable, and understandable.

The following describes the process of normalization:

1. Each entity is split (moved) out into its own table.
2. Each discrete attribute is created as its own column within the table.
3. Each entity instance (row) uses a primary key to identify it uniquely.
4. Related entities are linked with foreign key columns.

To conclude the previous steps in the normalization process, this data management technique organizes and represents data in a relational database more efficiently and consistently. By following the normalization steps, you can eliminate data redundancy, ensure data integrity, and simplify data manipulation and querying.

In *Figure 2.4*, you can see this process in action; the scenario you will explore is where there is a Sales data table with two rows containing duplicate entries for the order number and customer details; the Product column has an entry that is not a single-value character; it combines the attributes of name and price. You can change how the data is organized and represented in the database using normalization.

In *Figure 2.4*, you can see how the single Sales data table has been altered, so each data entity has been split into tables; duplicate data has been removed from the tables, such as the order number and customer details from the Sales data table, and each attribute of product name and price has been moved to its own column.

Sales data				
order_no	order_date	customer	product	quantity
500	6/1/24	steve miles, 1 data street, reading	SI1210 (£899)	2
500	6/1/24	steve miles, 1 data street, reading	Roland TB303 (£3,000)	1

Not Normalized Data

Normalized Data

Customer				
customer_id	first_name	last_name	address	city
smiles_70	steve	miles	1 data street	reading

Order		
order_id	order_date	customer_id
500	6/1/24	smiles_70
500	6/1/24	smiles_70

LineItem			
order_id	item_no	Item_id	Item_qty
500	1	1992	2
500	2	1988	1

Product		
item_id	item_name	item_price
1992	SI1210	3000.00
1988	Roland TB303	899.00

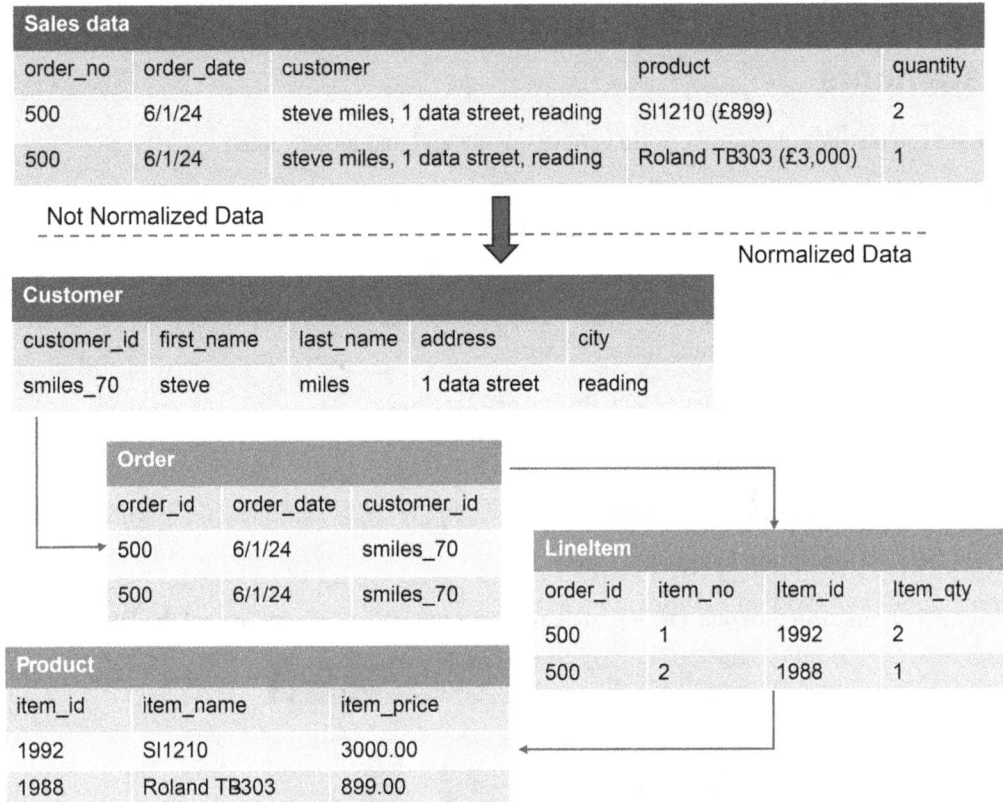

Figure 2.4 – Normalized data example

The benefit of normalization is that data manipulation is more efficient; in the scenario shown in *Figure 2.4*, should you wish to update the customer's address details, you only need to change a single row value. In contrast, with the first table of data in *Figure 2.4*, with no normalization, you would need to change each row.

In this section, you learned about normalization. The following section looks at the SQL statements that can be used to query and manipulate relational data.

Identify Common Structured Query Language (SQL) Statements

SQL statements allow users to perform various queries and manipulate data.

SQL statements are instructions for managing and inquiring about data in relational databases. They follow specific rules and syntax for data interaction.

SQL statements are crucial for those looking to work with data, such as conducting queries and making updates, deletions, or additions. They enable data professionals to handle data operations smoothly and proficiently, which is essential in various sectors.

There are three logical categories of SQL statements, as follows:

- **Data Control Language (DCL)**
- **Data Definition Language (DDL)**
- **Data Manipulation Language (DML)**

Figure 2.5 summarizes the common SQL statements for the *Skills Measured* area of the *DP-900* exam:

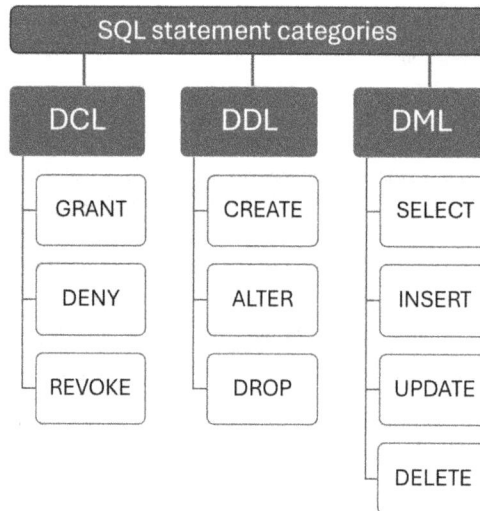

Figure 2.5 – Common SQL statements

In the following sections, you will learn more about these SQL statement categories and some examples to help you understand them.

DCL Statements

Database access is managed with DCL statements. These represent who can *access, update,* or *delete* objects in a database.

DCL statements include GRANT, REVOKE, and DENY.

GRANT

This statement permits specific user or role actions on database objects.

This could be permissions such as SELECT, INSERT, UPDATE, DELETE, and EXECUTE on database objects such as tables, views, and stored procedures. The GRANT statement allows for better control over allowing users access/security to database objects, thereby granting only the specified access type on the defined database objects.

The following is an example of a GRANT statement:

```
GRANT SELECT ON Product TO example_user;
```

DENY

This statement is used to explicitly refuse one or more permissions to a user or a role on an object within the database, such as a table, view, or procedure. This statement does not allow any permissions beyond this single DENY statement for the specified object.

Users may have been given access to permissions through a GRANT statement, and the access granted might be equal to or greater than the right being denied. However, a DENY statement applies restrictions on top of any granted access. This statement is useful for setting a security policy to stop certain actions for a given user or role, regardless of any other granted permissions.

The following is an example of a DENY statement:

```
DENY SELECT ON Product TO example_user;
```

REVOKE

This statement moves previously granted permissions from specific users or roles.

The privileges you granted to the user are revoked, so the user can no longer access the data or feature. The REVOKE statement is critical to database security. It's the mechanism by which access is controlled; without it, data and features irrelevant to a user would be exposed, and data not intended for them could be changed.

The following is an example of a REVOKE statement:

```
REVOKE SELECT ON Product FROM example_user;
```

This will allow database administrators to restrict users' access to sensitive information when running the statements.

DDL Statements

DDL statements act **on the structure of a database.**

DDL statements include CREATE, ALTER, and DROP.

CREATE

This statement creates a new database, a new table in an existing database, or database objects such as triggers, views, functions, and so on.

To create a table as an example, you specify the table's name and define the columns and their data types and constraints. It allows the database structure to be defined, such as the fields used to store the data. It is a core statement in database management.

The following is an example of a CREATE statement:

```
CREATE TABLE Employees (
    EmployeeID INT PRIMARY KEY,
    FirstName VARCHAR(50),
    LastName VARCHAR(50),
    Department VARCHAR(50),
    Salary DECIMAL(10, 2)
);
```

ALTER

This statement allows you to change the structure of an existing database object, such as a table, view, or index, on the fly without having to drop and recreate it. This operation includes adding, altering, or deleting a column from a table, changing the data types, renaming objects, or defining constraints. It is often needed to maintain database schemas when changes to requirements occur without losing the existing data.

The following is an example of an ALTER statement:

```
ALTER TABLE Employees
ADD COLUMN Email VARCHAR(100);
```

DROP

This statement removes a database object (e.g., a table, an index, a view, or a database) already in the database. Once this statement is executed, the object and all associated data, structures, and dependencies are removed from the system. Being destructive, the statement use of DROP is carefully controlled, typically granting it only to users with higher-level administrative privileges to prevent a user from accidentally deleting data.

The following is an example of a DROP statement:

```
DROP TABLE Employees;
```

Using DDL statements, databases can be optimized for performance and scalability.

DML Statements

DML statements are used to *access* data stored in a database.

DML statements include SELECT, INSERT, UPDATE, and DELETE.

SELECT

This statement queries and returns results from a database. In the SELECT clause, you can return the columns of interest, and in the FROM clause, the table from which to return them.

Results can be filtered using the WHERE clause and ordered using the ORDER BY clause. Results can be grouped using the GROUP BY clause and filtered by groups using the HAVING clause. Joins can be marked using clauses such as JOIN across multiple tables to combine rows. The SELECT statement is at the core of SQL.

The following is an example of a SELECT statement:

```
SELECT *
FROM Employees
WHERE Department = 'Engineering';
SELECT ProductID, Name, Price
FROM Product
ORDER BY Price ASC;
```

INSERT

This statement is used to insert new data rows into a database table. It specifies the table where the data should be inserted and the values entered into each table column. The syntax is for the INSERT INTO clause, the name of the table we are inserting data into, an optional list of columns, and the VALUES keyword, followed by a list of values corresponding to the columns we specified. The INSERT statement can be used with the SELECT statement to insert data from another table.

The following is an example of an INSERT statement:

```
INSERT INTO Employees (EmployeeID, Title, LastName, Department)
VALUES (1, 'Mr', 'blogs', 'IT')
       (2, 'Mrs', 'Miggins', 'RETAIL')
```

UPDATE

This statement is used to update one or more columns for rows in a table that is a part of a database.

This means it can be used to amend your database content so it matches the records you consider the most correct. With this statement, it is possible to edit one or more columns for all the rows of a table that match a certain condition. The syntax generally contains the table name, the SET clause that includes the columns to be changed and their updated values, and the WHERE condition.

This statement is important for maintaining data integrity and accuracy.

The following is an example of an UPDATE statement:

```
UPDATE Employees
SET Department = 'Operations'
WHERE EmployeeID = 1;
```

DELETE

This statement removes one or more rows from a table by evaluating certain conditions.

It has a simple syntax where the table name from which records need to be removed should be specified first. Optionally, a WHERE statement can specify the conditions determining which rows need to be deleted. DELETE statements should be used with caution as they delete data permanently, and without a WHERE clause, they can permanently lose all data.

The following is an example of a DELETE statement:

```
DELETE FROM Employees
WHERE EmployeeID = 1;
```

Users can retrieve and modify the data they need from a database using DML statements.

Now that you have progressed your relational database knowledge with some common SQL statement skills, you will learn how to identify some common database objects.

Identify Common Database Objects

The content in this section will help you learn about some common database objects that are critical to know about, such as tables, views, indexes, and stored procedures.

A fundamental understanding of the purpose of a table, a view, an index, and a stored procedure helps manage data efficiently, improve performance and security, and attain other goals a database system might have.

SQL **database objects** are entities or structures within a database that help to organize and store data in a structured manner. Examples of SQL database objects included in the *DP-900* exam scope are as follows:

- **Tables** are the containers that store the actual data
- **Views** are windows into the data, providing a simplified, access-controlled, secured view
- **Indexes** are maps that help speed the retrieval of data
- **Stored procedures** are code containers that can enclose a set of interrelated operations for portable reuse and consistency

You will now look at each of these objects in the following sections.

Tables

Relational Database Management Systems (**RDBMSs**) store and retrieve data from tables, providing a simple but powerful and organized method of storing large amounts of data for most database-driven applications.

Tables are the most fundamental and common database objects. Data is contained in rows and columns; a table contains as many rows as necessary (to capture as much data as needed).

Each row of a table is an instance (a record) of data, and each column of that record represents an attribute of the instance.

Tables are created and modified using SQL commands, and queries are made using SQL statements to retrieve and manipulate datasets.

Tables are interrelated through key fields, and data can be queried across all tables related to a specific table using a specific key.

An example SQL `CREATE TABLE` command would be as follows:

```
CREATE TABLE Employees (
    EmployeeID INT PRIMARY KEY,
    FirstName VARCHAR(50),
    LastName VARCHAR(50),
    Department VARCHAR(50),
    Salary DECIMAL(10, 2)
);
```

Views

Views provide a way to customize and manage data visibility in a database. They are virtual tables that display data from one or more tables; these are considered predefined SQL queries, which typically means SQL queries that are written and stored in the database ahead of time for specific tasks or functions.

Unlike physical tables, views do not store any data themselves but, instead, present data from the underlying tables in an easier way to work with or understand. Views are created using SQL SELECT queries that specify the fields and conditions to include in the view.

An example SQL CREATE VIEW command would be as follows:

```
CREATE VIEW EmployeeDetails
AS
SELECT EmployeeID, FirstName, LastName, Salary
FROM Employees;
```

You can query the EmployeeDetails view created using the following statement:

```
SELECT *
FROM EmployeeDetails;
```

Once created, views can be queried and manipulated just like physical tables, allowing users to retrieve, update, insert, and delete data as needed.

Views can be used when presenting data in a specific format or masking sensitive or confidential data from certain users. They can also simplify complex queries or provide a simplified interface to data for non-technical users.

Indexes

Indexes improve the performance of database queries; they form **tree-based** structures.

Tree-based structures are data structures that organize data in a hierarchical way, where each node has a parent node and zero or more child nodes. Tree-based structures can store data with a natural hierarchy, such as files and folders, categories and subcategories, or family trees.

Indexes are data structures used to improve the speed and efficiency of database queries. They work by creating a searchable index of the data in a table or set of tables, which allows the database to quickly find and retrieve specific data without scanning the entire table.

An example SQL CREATE INDEX command would be as follows:

```
CREATE INDEX idx_salary ON Employees(Salary);
```

One or more columns in a table can have indexes created and can be either **clustered** or **non-clustered**.

Clustered indexes determine the *physical order* of data in a table. **Non-clustered indexes** consist of index keys pointing to the *data's location* (row) in the table. Database queries can be executed much faster using indexes, particularly for large tables with many records. However, indexes can also have a downside – they can increase the time required to insert, update, or delete data from a table. As a result, it is important to use indexes judiciously and carefully consider the trade-offs between query performance and data modification performance when designing database schemas.

Stored Procedures

These are **precompiled sets** of SQL statements with parameters that can be executed multiple times. By grouping SQL statements in a stored procedure, a developer can create a single unit of code (which is reusable) that can be called from various places within a database or application. This can help simplify application development, reduce the amount of code to be created, and make it more straightforward to maintain and update complex database operations over time. One of the key benefits of using stored procedures is that they can encapsulate complex business logic and implement sophisticated data processing workflows.

An example SQL CREATE PROCEDURE command would be as follows:

```
CREATE PROCEDURE GetEmployeeByID(IN emp_id INT)
BEGIN
    SELECT *
    FROM Employees
    WHERE EmployeeID = emp_id;
END;
```

The command used to execute this stored procedure would be as follows:

```
EXECUTE GetEmployeeByID(1);
```

Stored procedures can also be used to improve the performance of database operations by reducing the amount of data that needs to be transferred between the database and the application. By performing data processing tasks within the database server rather than in the client application, stored procedures can significantly reduce the amount of network traffic and improve the overall efficiency of database operations.

This section on the common relational database objects concludes this part for Azure. This section also concludes the learning content for this chapter. Now, it's time to summarize what skills you have learned in this chapter.

Summary

In this chapter, you learned how to describe relational concepts essential for working with Azure data services.

You explored how to define relational data, normalization, and objects such as tables, views, indexes, and stored procedures in relational databases. Then, you learned about the SQL language and how it is used to query and manipulate data in relational databases.

By mastering these concepts, you have gained a solid foundation for understanding and working with relational data in Azure and the skills you need to succeed in the Azure Data Fundamentals exam confidently. You are ready to move on to the next chapter, where you will learn about non-relational concepts and how they apply to Azure data services.

This chapter included complete coverage of the *DP-900 Azure Data Fundamentals* exam *Skills Measured* area, *Describe Relational Concepts*.

The next chapter will cover **relational Azure data services** to follow the *Skills Measured* sequence in Microsoft's study guide for the *DP-900* exam.

Exam Readiness Drill – Chapter Review Questions

Apart from a solid understanding of key concepts, being able to think quickly under time pressure is a skill that will help you ace your certification exam. That is why working on these skills early on in your learning journey is key.

Chapter review questions are designed to improve your test-taking skills progressively with each chapter you learn and review your understanding of key concepts in the chapter at the same time. You'll find these at the end of each chapter.

> **How to Access These Materials**
>
> To learn how to access these resources, head over to the chapter titled *Chapter 9, Accessing the Online Resources.*

To open the Chapter Review Questions for this chapter, perform the following steps:

1. Click the link – `https://packt.link/DP900Ch02`

    Alternatively, you can scan the following **QR code** (*Figure 2.6*):

Figure 2.6 – QR code that opens Chapter Review Questions for logged-in users

2. Once you log in, you'll see a page similar to the one shown in *Figure 2.7*:

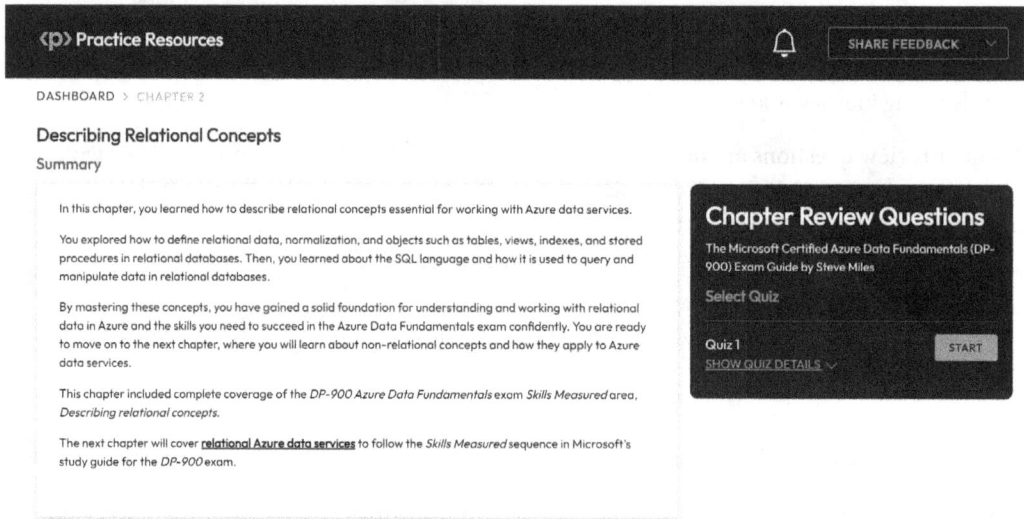

‹p› Practice Resources   🔔   SHARE FEEDBACK ⌄

DASHBOARD > CHAPTER 2

Describing Relational Concepts
Summary

In this chapter, you learned how to describe relational concepts essential for working with Azure data services.

You explored how to define relational data, normalization, and objects such as tables, views, indexes, and stored procedures in relational databases. Then, you learned about the SQL language and how it is used to query and manipulate data in relational databases.

By mastering these concepts, you have gained a solid foundation for understanding and working with relational data in Azure and the skills you need to succeed in the Azure Data Fundamentals exam confidently. You are ready to move on to the next chapter, where you will learn about non-relational concepts and how they apply to Azure data services.

This chapter included complete coverage of the *DP-900 Azure Data Fundamentals* exam *Skills Measured* area, *Describing relational concepts.*

The next chapter will cover **relational Azure data services** to follow the *Skills Measured* sequence in Microsoft's study guide for the *DP-900* exam.

Chapter Review Questions
The Microsoft Certified Azure Data Fundamentals (DP-900) Exam Guide by Steve Miles

Select Quiz

Quiz 1   START
SHOW QUIZ DETAILS ⌄

Figure 2.7 – Chapter Review Questions for Chapter 2

3. Once ready, start the following practice drills, re-attempting the quiz multiple times.

Exam Readiness Drill

For the first three attempts, don't worry about the time limit.

ATTEMPT 1

The first time, aim for at least **40%**. Look at the answers you got wrong and read the relevant sections in the chapter again to fix your learning gaps.

ATTEMPT 2

The second time, aim for at least **60%**. Look at the answers you got wrong and read the relevant sections in the chapter again to fix any remaining learning gaps.

ATTEMPT 3

The third time, aim for at least **75%**. Once you score 75% or more, you start working on your timing.

Tip

You may take more than **three** attempts to reach 75%. That's okay. Just review the relevant sections in the chapter till you get there.

Working On Timing

Target: Your aim is to keep the score the same while trying to answer these questions as quickly as possible. Here's an example of how your next attempts should look like:

Attempt	Score	Time Taken
Attempt 5	77%	21 mins 30 seconds
Attempt 6	78%	18 mins 34 seconds
Attempt 7	76%	14 mins 44 seconds

Table 2.1 – Sample timing practice drills on the online platform

---

Note

The time limits shown in the above table are just examples. Set your own time limits with each attempt based on the time limit of the quiz on the website.

With each new attempt, your score should stay above **75%** while your "time taken" to complete should "decrease". Repeat as many attempts as you want till you feel confident dealing with the time pressure.

3

Describe Relational Azure Data Services

Imagine you are managing the backend for a large e-commerce website with thousands of transactions per second and various complex queries for reporting and analytics. Relational databases are at the core of such scenarios. The database management system provides a structured way to store, manipulate, and retrieve (query) data.

This chapter discusses the Azure relational data services, a key *DP-900 Azure Data Fundamentals* exam *Skills Measured* topic. These data services in an Azure environment are one of the core building blocks for a cloud-enabled solution.

Azure offers a suite of relational data services designed to cater to various needs, from traditional on-premises database migrations to modern, cloud-native applications.

Your options for Azure environment data services include implementing Microsoft and open source relational database systems. Understanding these options is essential for passing the exam and applying these choices in real-life scenarios, where selecting the most applicable data service for a given scenario is key.

In this chapter, you will cover two key areas:

- Microsoft databases on Azure
- Open source databases in Azure

The topics are related to the *DP-900 Azure Data Fundamentals* certification exam, covering measured skills for identifying relational data considerations on Azure; this corresponds to about 20–25% of the total measured skills. Completing this chapter ensures you have the skills to identify relational data services on Azure and are ready for the certification exam.

Now, you can look at the different methods available for deploying Microsoft SQL databases within Azure environments in the following section.

Microsoft SQL Databases on Azure

In this chapter, you will explore the **Azure SQL family**, covering **Azure SQL Database**, **Azure SQL Managed Instance**, and **SQL Server on Azure Virtual Machines** (**VMs**).

These products offer varied features for deploying and managing SQL databases in Azure environments, catering to diverse business needs.

Microsoft SQL databases use the **Online Transaction Processing** (**OLTP**) data processing model covered in *Chapter 1, Describe Core Data Concepts*.

The available Azure deployments for the Microsoft SQL **Relational Database Management System** (**RDBMS**) are visually represented in *Figure 3.1*:

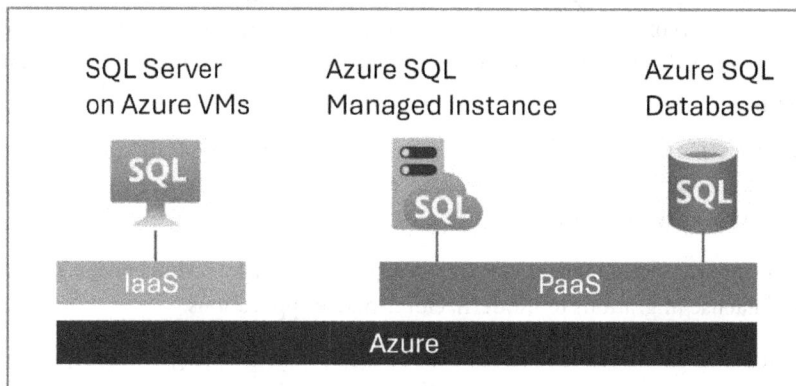

Figure 3.1: Azure Microsoft SQL RDBMs deployments

You must understand the deployments for Microsoft SQL RDBMS since they are outlined in the *Identify Considerations for Relational Data on Azure Skills Measured* area.

The following provides a brief outline of each Microsoft SQL RDBMS deployment:

- **SQL Server on Azure VMs**: A fully customer-managed **infrastructure-as-a-service** (**IaaS**) relational database service that allows users to access the full SQL Server engine and related services on IaaS Azure VMs. The advantage of this deployment method is that 100% compatibility with "on-premises" deployments is guaranteed; it is an excellent choice for "lift-and-shift" scenarios, where on-premises workloads are moved to Azure regions without requiring code or configuration changes to apps or dependent workloads. This deployment method provides full access and control of the SQL Server instance, the underlying **operating system** (**OS**), and compute resources.

- **Azure SQL Managed Instance**: A fully Microsoft-managed **platform-as-a-service** (PaaS) relational database service that abstracts the underlying compute resources and OS. Azure SQL Managed Instance can support multiple databases. It provides nearly 100% compatibility with the latest SQL Server database engine, is well suited for "lift-and-shift" scenarios, and allows modernization of existing applications with minimal code or configuration changes. Automated updates, backups, and other operational tasks are provided to reduce the maintenance and administration overhead.

- **Azure SQL Database**: A fully Microsoft-managed PaaS relational database service that abstracts the underlying SQL Server instance. It offers database instances on a fully managed SQL Server database engine, offering automated high availability, scalability, backups, patching, and monitoring with minimal management overhead. Modernization and new cloud-first workloads are best suited for this deployment method.

- **Azure SQL Edge**: A data engine for real-time analytics and machine learning on data from IoT devices and edge sources. It is designed for scenarios where data needs to be processed and analyzed quickly at edge locations without constant public cloud connectivity. **Edge locations** refer to the data centers and points of presence that are situated closer to the end users; these edge locations help reduce the latency and improve the overall performance by caching the content closer to the user's geographical location.

- This deployment method is ideal for predictive maintenance, real-time monitoring, and remote asset management.

> **Note**
>
> For completeness, Azure SQL Edge is included in this list but not detailed further in this chapter; you can read more info at `https://learn.microsoft.com/en-GB/azure/azure-sql-edge/`.

Microsoft manages the underlying compute, OS, and SQL Server instance with the Azure SQL PaaS databases, while with IaaS, the customer manages everything.

SQL Server Management Studio (SSMS) and **Azure Data Studio** are **graphical user interface (GUI)** tools for database tasks. **Azure Data Studio** can detect database deployment compatibility issues with the **Azure SQL Migration extension**. A command-line utility that can be used is **sqlcmd**.

In this context of Microsoft SQL databases on Azure, deployment refers to how the SQL database platform is set up, managed, and accessed within the Azure environment. Different deployment methods provide various levels of control, flexibility, and automation, depending on the needs of the business. Each method offers distinct features, making it suitable for different scenarios, from fully managed services to more flexible customer-controlled and managed solutions.

All deployment methods provide native **virtual network** (**VNet**) integration; this allows network segmentation and isolation from your databases. Using **Azure Private Link** means public access is not required; access to your databases will be over a private endpoint in the VNet. Azure SQL Database provides network-level access control to restrict external access through IP firewall rules.

The first Microsoft SQL database on Azure deployment method you will explorer in more detail will use Azure VMs as a compute and storage platform for Microsoft SQL Server.

SQL Server on Azure VMs

SQL Server on Azure VMs is an IaaS deployment option for **Microsoft SQL**, which is a fully supported way to quickly "lift and shift" your on-premises SQL Server deployments to run in an Azure region with 100% compatibility. It requires no configuration or code changes to any workloads that make connections to the databases, or workloads where you still need control and access to SQL Server instance-level features and the underlying OS.

You can quickly provision VMs with pre-configured SQL Server images or create custom images; you maintain complete control over the SQL Server version, edition, OS, and VM type and sizing. Microsoft will provide an availability of 99.99%.

As an IaaS service, you are responsible for managing all aspects of the SQL Server instance and underlying OS and VM instance; this includes deployment, configuration, security, backup, patching, upgrading, scaling, performance monitoring, and so on.

Now, you will compare what you have learned about the IaaS service SQL Server on Azure VMs with the PaaS service Azure SQL Managed Instance.

Azure SQL Managed Instance

Azure SQL Managed Instance is a fully managed PaaS deployment method for Microsoft SQL.

Full SQL server access is provided, including a patched OS with 99.99% built-in high availability. Native VNet support is provided for network segmentation and access control.

It is designed for customers migrating apps from an on-premises or IaaS environment to a fully managed PaaS environment, where you no longer need control or access to the underlying OS. Microsoft provides automatic patching, version updates, automated backups, and high availability.

Azure SQL Managed Instance offers a native VNet implementation that addresses common security concerns for PaaS services. This ensures that your database is isolated within your VNet.

Purchasing Models

With Azure SQL Managed Instance, a **vCore-based** purchasing model is offered, and it is the recommended purchasing model type for workloads that will move to Azure. This model allows the simplest translation of on-premises resource allocation to Azure-hosted workloads.

With this purchasing model, you can independently select the amount of compute and storage to meet your requirements. This approach is visualized in *Figure 3.2*:

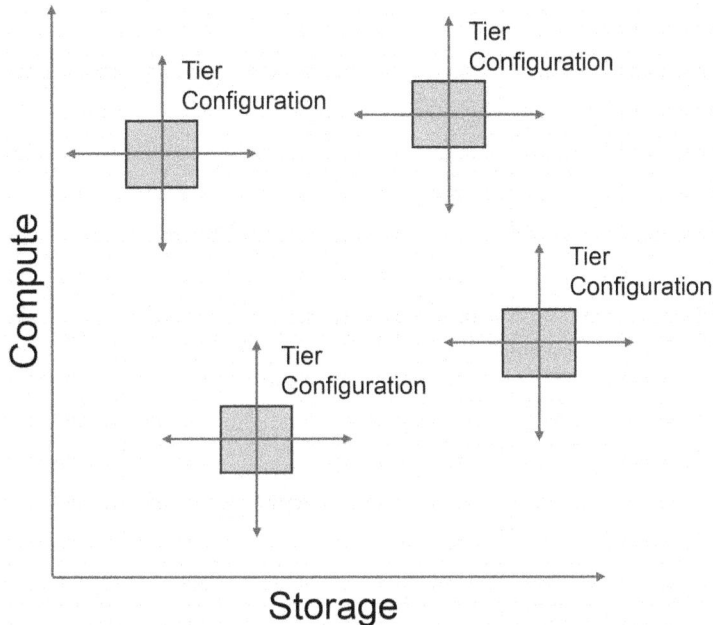

Figure 3.2: The vCore purchasing model

In *Figure 3.2*, each square with arrows represents a different configuration or tier in the vCore purchasing model for Azure SQL databases; depending on the specific configurations available in your context, you can label each square according to different service tiers or configurations (e.g., "General Purpose," "Business Critical," "Hyperscale," and so on).

You can also leverage **Azure Hybrid Benefit for SQL Server** for your Azure SQL Managed Instances and Azure SQL databases; this allows you to use any existing eligible SQL license you own instead of being billed for Microsoft-provided SQL licenses.

Service Tiers

Service tiers are collated features and capabilities based on performance and business continuity specifications to meet your workload requirements. The **vCore** model provides the following two service tiers:

- **General-purpose**: This is intended for typical application performance and requirements of I/O latency

- **Business-critical**: This is intended for low I/O requirements applications, and maintenance operations of the underlying platform have minimal workload impact

Next, you will look at the PaaS service of Azure SQL Database.

Azure SQL Database

Azure SQL Database is a PaaS deployment option for Microsoft SQL that also provides a scalable, fully managed database engine service. It is best suited for modern cloud-based and new cloud-native applications.

As a PaaS delivery model, it simplifies provisioning, patching, upgrading, and monitoring SQL and the underlying OS and backup. Azure SQL Database offers a robust SLA for availability and uptime of 99.995%.

Database management aspects of updates, backup, and recovery are fully automated and managed by Microsoft as part of the fully managed service; this also extends to the underlying OS for the managed instance of SQL Server.

Deployment Methods

Azure SQL Database is provided in the following two deployment methods:

- **Single database**: This is a standalone database instance provisioned on the Microsoft-managed SQL Server instance; it operates independently with dedicated resources and management controls tailored to its specific needs. It suits scenarios where each application or workload requires a separate isolated database instance with predictable performance and high resource utilization.

- **Elastic pool**: This is a collection of databases provisioned on the Microsoft-managed SQL Server instance but with a shared set of resources, where the resources are dynamically allocated among the databases based on demand; it is the best option where multiple databases have varying and unpredictable usage patterns.

In this section, you explored the two deployment methods available for Azure SQL Database: single database and elastic pool.

Understanding these deployment methods is a measured *DP-900 exam* skill and, in the real world, crucial for effectively managing SQL databases on Azure, allowing you to choose the right approach based on specific application requirements and workload patterns.

Purchasing Models and Service Tiers

With Azure SQL Database, there are two purchasing models offered:

- **vCore-based**; the amount of computing and storage can be independently selected (refer to the previous *Azure SQL Managed Instance* section to recap on this purchasing model).

  - The **vCore** model is also available with the following three service tiers:

- **General-purpose**: This is intended for typical workloads; it has balanced options for **compute** and **storage** that are cost-conscious.

- **Business-critical**: This is intended for OLTP workloads that require low latency I/O and high transactions. Isolated replicas provide the highest failure resilience.

- **Hyperscale**: This is intended for most business workloads that require scale and storage resources that can be scaled independently. Isolated database replicas can be configured.

- **DTU-based** selection of the processing, memory, and I/O resources are bundled and pre-defined into different mixes for different compute sizes to select from to support different database workloads. This approach is visualized in *Figure 3.3*:

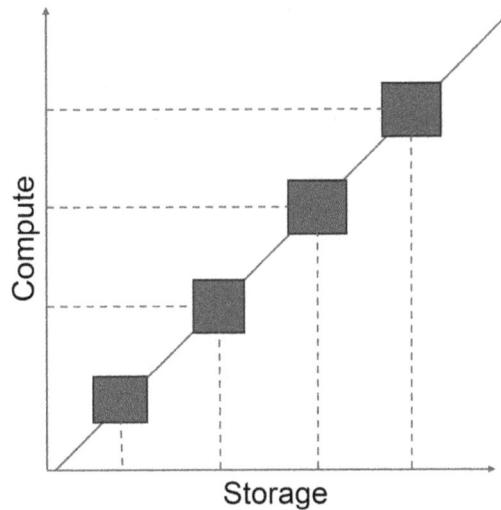

Figure 3.3: DTU purchasing model

The **DTU** model provides the following two service tiers:

- **Standard**: This is intended for typical workloads; it has balanced options for compute and storage that are cost-conscious.

- **Premium**: This is intended for **OLTP** workloads that require low latency I/O and high transactions. Isolated replicas provide the highest failure resilience.

Security

Security of Azure SQL databases for external access (not within the Azure network boundary) can be provided through network access controls through IP firewall rules.

By default, there is no external access to your database until a firewall rule has been created to explicitly allow access from a specific IP address or range of IP addresses.

Should the IP address of the client accessing your database vary, you must update the firewall settings to accommodate this new IP address; this administration task creates a management overhead.

You can also configure **private access**, which restricts access and can be set to allow access only from within a specific Azure boundary VNet or via the **Private Link** service, which permits access only from a **private endpoint** within an Azure boundary VNet.

In the final section on *Microsoft SQL databases on Azure*, you will compare the deployment options covered.

Comparing SQL Deployment Options

As mentioned earlier in this chapter, in the context of Microsoft SQL databases on Azure, SQL deployment refers to setting up, managing, and accessing the SQL database platform in the Azure environment. This involves choosing the platform or infrastructure service approach, adjusting settings, and deciding the level of control and automation needed for the business. Different deployment options provide various levels of management, control, and flexibility to meet different business needs.

Each Microsoft SQL deployment option offers different control and management overhead levels, features, and scalability models catering to various application requirements and preferences; each deployment is compared in this section.

Figure 3.4 outlines the different Microsoft SQL deployment options:

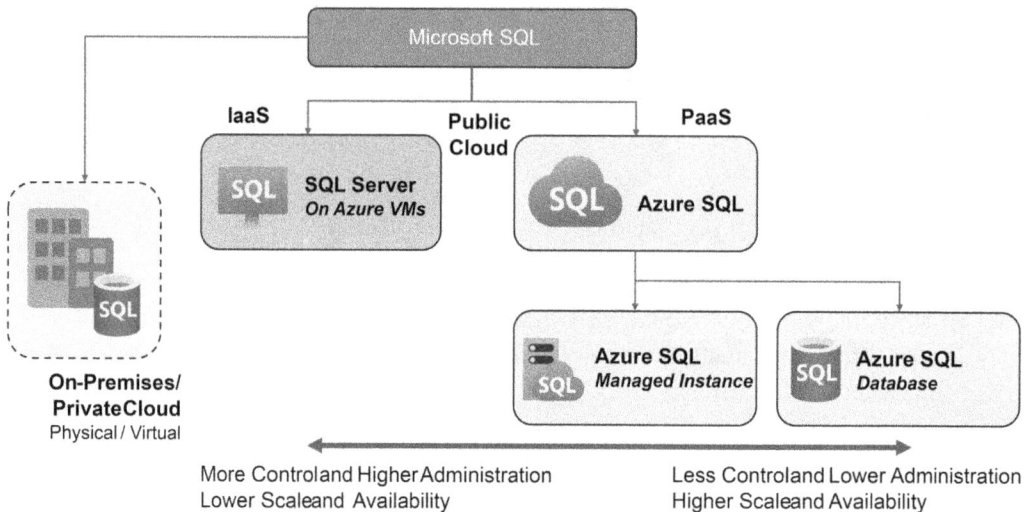

Figure 3.4: SQL deployment options

The Azure deployment methods can be summarized as follows:

- **SQL Server on Azure VMs**:

  - It is suited for most "lift-and-shift" migrations of on-premises instances of SQL servers and workloads where access to the OS is required

  - 100% compatibility with SQL Server

  - The customer is responsible for installing the SQL components, such as the SQL engine, and all updates for the SQL instance and the OS

  - Backups must be performed manually by the customer

  - Scale and high availability must be manually designed into the environment

  - Protection through **vulnerability assessments** and **advanced threat protection**, such as that provided by Microsoft Defender for SQL, is not enabled

  - SQL instance cannot be paused to reduce SQL costs

- **Azure SQL Managed Instance**:

  - It is suited for most "lift-and-shift" migrations of on-premises instances of SQL servers and workloads, where access to the OS is not required or there is a wish to reduce admin overhead

  - Near 100% compatibility with SQL Server; instance-level features, such as Database Mail, cross-database queries, and transactions, will be supported

  - Microsoft is responsible for installing the SQL components, such as the SQL engine and all updates for the SQL instance and the OS; this means the latest features will always be available

  - Microsoft automatically completes backups

  - Scale and high availability are built into Microsoft's service

  - Microsoft Defender protects vulnerability assessments and advanced threat protection for SQL through a **Microsoft Defender for Cloud** Defender plan

  - SQL instance cannot be paused to reduce SQL costs

- **Azure SQL Database**:

  - It is suited for modern cloud workloads, with no SQL instance management overhead

  - Microsoft is responsible for installing the SQL components, such as the SQL engine and all updates for the SQL instance and the OS; this means the latest features will always be available

  - Microsoft automatically completes backups

  - Scale and high availability are built into Microsoft's service

- Microsoft Defender protects vulnerability assessments and advanced threat protection for SQL through a Microsoft Defender for Cloud plan

- SQL Azure Database Serverless can be paused to reduce SQL costs

The choice also depends on the level of control and management needed. The following illustration in *Figure 3.5* intends to aid in visualizing this:

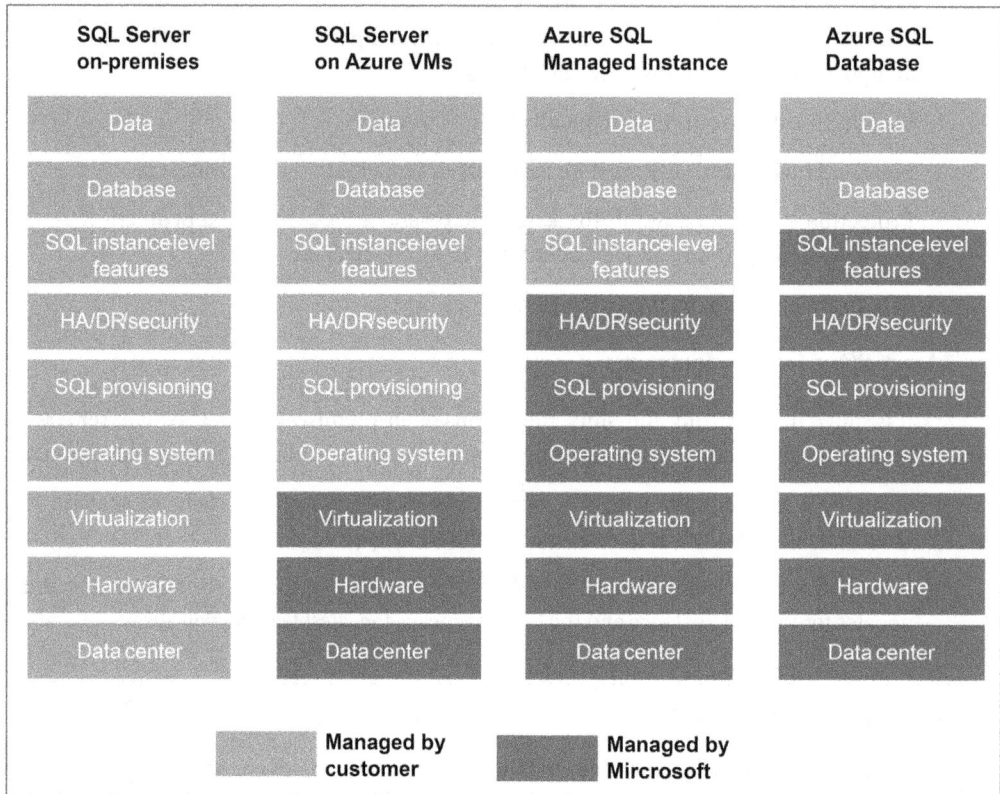

SQL Server on-premises	SQL Server on Azure VMs	Azure SQL Managed Instance	Azure SQL Database
Data	Data	Data	Data
Database	Database	Database	Database
SQL instance-level features	SQL instance-level features	SQL instance-level features	SQL instance-level features
HA/DR/security	HA/DR/security	HA/DR/security	HA/DR/security
SQL provisioning	SQL provisioning	SQL provisioning	SQL provisioning
Operating system	Operating system	Operating system	Operating system
Virtualization	Virtualization	Virtualization	Virtualization
Hardware	Hardware	Hardware	Hardware
Data center	Data center	Data center	Data center

Managed by customer   Managed by Mircrosoft

Figure 3.5: SQL deployment options responsibility and control model

In *Figure 3.5*, you can see the areas that you will be responsible for and control and those that Microsoft will be responsible for and control.

Now that you have learned the available options for implementing Microsoft SQL databases on Azure, the next section will cover the available options on Azure for implementing open source databases.

Open Source Databases on Azure

The integration of cloud technology in business operations continues to drive the demand for agile, reliable, scalable, and cost-efficient database solutions. With their flexibility, innovation, and feature-rich functionality, open source databases can be a great choice for many scenarios. This section will look at deploying and managing **MySQL**, **PostgreSQL**, and **MariaDB** on Azure.

First, we'll cover **Azure Database for MySQL**, a fully managed PaaS service that combines the strengths of MySQL with the scalability and reliability of Azure; then, we'll spend time discussing **Azure Database for PostgreSQL**, a powerful and flexible open source relational database that is compatible with many other systems; finally, we'll look at **Azure Database for MariaDB**, a high-performing, highly scalable open source relational database that is simple to configure and use.

For a business, it is important to know that these database selection options mean that you can focus on innovation and take advantage of Azure's many managed services, backed by enterprise-class security, integrity, and availability.

By the end of the chapter, you'll know which open source databases are available on Azure, how to deploy them, and why they're an important part of the modern data landscape.

You must have an understanding of the following databases and their deployment options since they are outlined in the *Identify Considerations for Relational Data on Azure* skills measured area:

- **MySQL**: An open source RDBMS that is widely used for web applications and known for its ease of use, scalability, and performance; it is mainly used with the **Linux, Apache, MySQL, and PHP (LAMP)** stack.

- **PostgreSQL**: An open source RDBMS known for its robustness, extensibility, and adherence to SQL standards, suitable for a wide range of applications; another function of PostgreSQL is to store geometric data, such as polygons.

- **MariaDB**: An open source relational RDBMS that is a fork of MySQL, designed for high performance, reliability, and ease of use; MariaDB is better suited for scenarios requiring querying data as it appeared at some point in the past. This is because a table in MariaDB can hold several versions of data.

With Azure, you can quickly provision and scale open source databases to meet your application requirements and pay only for what you use on a consumption model.

Figure 3.6 aims to represent the available Azure deployments for the open source RDBMS:

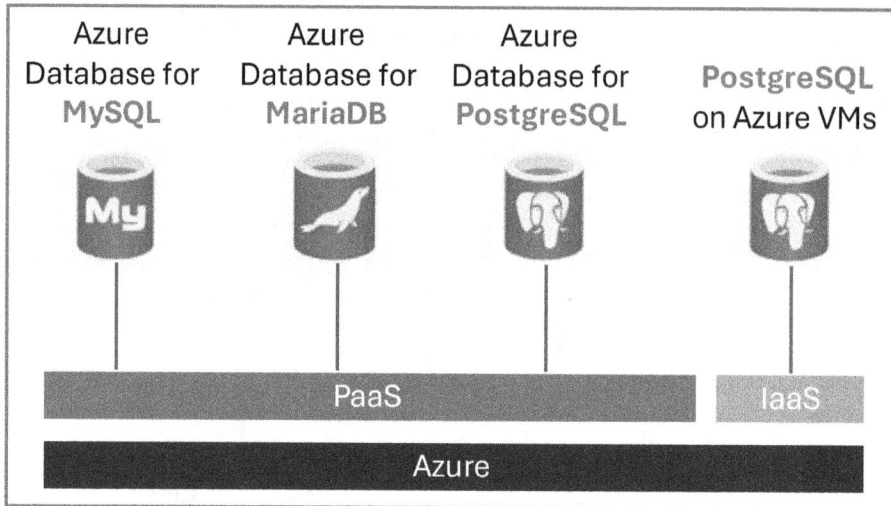

Figure 3.6: Azure open source RDBMS deployment options

Azure fully manages PaaS databases and offers enterprise-grade features, such as high availability, automatic backups, and security. PostgreSQL can also be deployed as IaaS, which allows for the control of the database software installed on an Azure VM.

The first open source database on Azure you will learn about will be MySQL.

MySQL Database

Azure Database for MySQL is a PaaS deployment option that provides a fully managed relational database service offered by Microsoft Azure, popular in LAMP deployment stacks.

It is a cloud-based service that provides a **MySQL database engine** based on the Community Edition of MySQL. Being a PaaS service, it benefits from predictable performance, built-in high availability, updates, automatic backups, and scaling capabilities. In addition to this, it is highly reliable and can handle many concurrent users and transactions; it also provides advanced enterprise-level security features to ensure legislation, compliance, and network security with firewall rules, providing encryption for data at rest and optional use of SSL connections for data in transit to help secure users' data. Data is protected through automatic backups that restore point-in-time data for the last 35 days.

The next Azure-hosted open source database that you will learn about will be PostgreSQL.

PostgreSQL Database

Azure Database for PostgreSQL is a PaaS deployment option that offers a fully managed relational database service from Microsoft Azure, built on the PostgreSQL Community Edition of the database engine.

Users can deploy, manage, and scale their PostgreSQL databases without worrying about server maintenance, patching, or backups.

Security is provided through network isolation, encryption at rest, and threat detection to help protect users' data. The database is managed with the **pgAdmin** tool.

You can monitor queries made on the database by querying the `query_store.qs_view` view for the database named `azure_sys`, which stores all the database queries created.

PostgreSQL is also available to be run in Azure regions, such as IaaS, with the database engine hosted on an Azure VM. This is a good option if you do not wish to hand over control of database operations to Microsoft; this allows you control of the database engine and better management for aspects such as configuration, patching, backups, and so on.

The final Azure-hosted open source database you will learn about is MariaDB.

MariaDB Database

Azure Database for MariaDB is a PaaS offering of MariaDB, made available as a fully managed relational database service from Microsoft Azure, built on top of the popular open source MariaDB community edition fork of MySQL. It is intended to provide a highly available, scalable, and secure database for cloud applications.

As a PaaS offering, it spares the need to manage the infrastructure and leaves you the core tasks of deploying and managing MariaDB databases. With Azure Database for MariaDB, you can easily create and manage relational databases, create automated backups, and monitor performance metrics. The data is protected through automatic backups that restore point-in-time for the last 35 days.

MySQL Workbench is a tool used to manage Azure Database for MariaDB. It simplifies database management tasks, such as creating and modifying objects, and provides monitoring tools to identify performance issues.

This section on the open source databases available on Azure concludes this section for the *DP-900 exam's Skills Measured* area, *Identify Considerations for Relational Data on Azure*.

This section also concludes the learning content for this chapter. Now, it is time to summarize the skills you learned in this chapter.

Summary

In this chapter, you learned how to describe the relational data services available for Azure environments, which provided essential skills needed to succeed confidently in the *Azure Data Fundamentals* exam.

This chapter included complete coverage of the *DP-900 Azure Data Fundamentals* exam's *Skills Measured* area, *Describe Relational Azure Data Services*.

The content emphasized themes such as understanding when to use each deployment option based on specific business needs and application requirements. It also showed that it is crucial to recognize the differences in management responsibilities between IaaS and PaaS services and how they impact operational overhead and resource allocation.

You are ready to move on to the next chapter, where you will learn about the topic of *Describe the Capabilities of Azure Storage* to follow the *Skills Measured* sequence in *Microsoft's study guide for the DP-900 exam*.

Additional Reading

This section provides links to additional exam information and study references:

- DP-900 - Microsoft Azure Data Fundamentals study guide: `https://learn.microsoft.com/en-us/credentials/certifications/resources/study-guides/dp-900`

- DP-900 - Microsoft Azure Data Fundamentals self-directed learning: `https://learn.microsoft.com/en-us/training/modules/explore-provision-deploy-relational-database-offerings-azure/`

- DTU-based purchasing model: `https://learn.microsoft.com/en-us/azure/azure-sql/database/service-tiers-dtu`

- vCore-based purchasing model: `https://learn.microsoft.com/en-us/azure/azure-sql/database/service-tiers-sql-database-vcore`

- Azure Hybrid Benefit for SQL Server: `https://azure.microsoft.com/pricing/hybrid-benefit/`

- What is Azure SQL Managed Instance?: `https://learn.microsoft.com/en-us/azure/azure-sql/managed-instance/sql-managed-instance-paas-overview`

- What is Azure SQL Database?: `https://learn.microsoft.com/en-gb/azure/azure-sql/database/sql-database-paas-overview`

Exam Readiness Drill – Chapter Review Questions

Apart from a solid understanding of key concepts, being able to think quickly under time pressure is a skill that will help you ace your certification exam. That is why working on these skills early on in your learning journey is key.

Chapter review questions are designed to improve your test-taking skills progressively with each chapter you learn and review your understanding of key concepts in the chapter at the same time. You'll find these at the end of each chapter.

> **How to Access These Materials**
>
> To learn how to access these resources, head over to the chapter titled *Chapter 9, Accessing the Online Resources.*

To open the Chapter Review Questions for this chapter, perform the following steps:

1. Click the link – `https://packt.link/DP900Ch03`.

   Alternatively, you can scan the following **QR code** (*Figure 3.7*):

Figure 3.7: QR code that opens Chapter Review Questions for logged-in users

2. Once you log in, you'll see a page similar to the one shown in *Figure 3.8*:

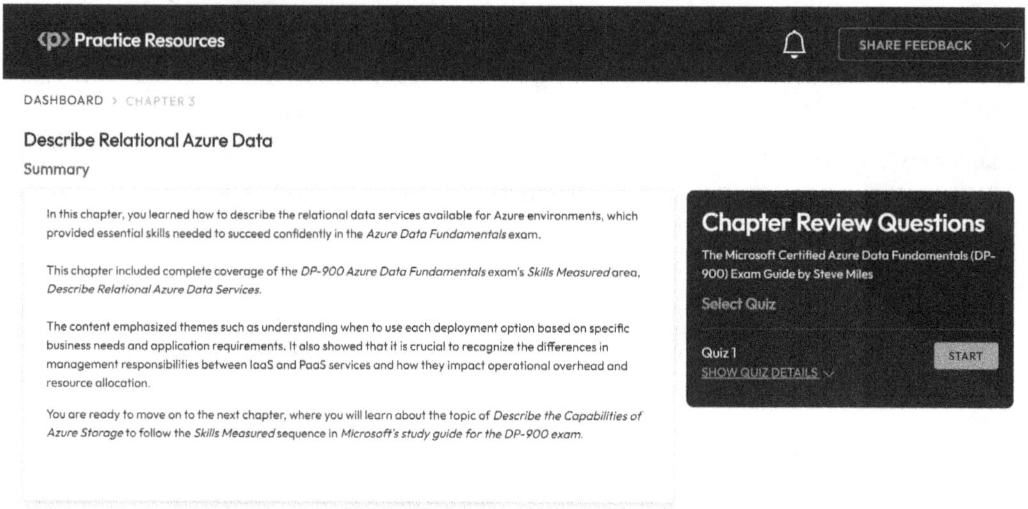

Chapter Review Questions

Figure 3.8: Chapter Review Questions for Chapter 3

3. Once ready, start the following practice drills, re-attempting the quiz multiple times.

Exam Readiness Drill

For the first three attempts, don't worry about the time limit.

ATTEMPT 1

The first time, aim for at least **40%**. Look at the answers you got wrong and read the relevant sections in the chapter again to fix your learning gaps.

ATTEMPT 2

The second time, aim for at least **60%**. Look at the answers you got wrong and read the relevant sections in the chapter again to fix any remaining learning gaps.

ATTEMPT 3

The third time, aim for at least **75%**. Once you score 75% or more, you start working on your timing.

> **Tip**
>
> You may take more than **three** attempts to reach 75%. That's okay. Just review the relevant sections in the chapter till you get there.

Working On Timing

Target: Your aim is to keep the score the same while trying to answer these questions as quickly as possible. Here's an example of how your next attempts should look like:

Attempt	Score	Time Taken
Attempt 5	77%	21 mins 30 seconds
Attempt 6	78%	18 mins 34 seconds
Attempt 7	76%	14 mins 44 seconds

Table 3.1: Sample timing practice drills on the online platform

> **Note**
>
> The time limits shown in the above table are just examples. Set your own time limits with each attempt based on the time limit of the quiz on the website.

With each new attempt, your score should stay above **75%** while your "time taken" to complete should "decrease". Repeat as many attempts as you want till you feel confident dealing with the time pressure.

4

Describe the Capabilities of Azure Storage

Consider a scenario where you are operating an online e-commerce business. As the customer base and sales grow, so does the data you need to keep track of – images of products, transaction records, recommendations, customer reviews, and much more. You must store this vast amount of rapidly generated data efficiently, ensure it is protected, and enable the correct level of access. This is what Azure Storage can provide; it is a robust, scalable, and secure service that helps take care of your data needs.

In this chapter, you will explore the capabilities of Azure Storage, an essential component of Microsoft's cloud platform, designed to handle a wide variety of data storage scenarios. It is also an essential skill for the DP-900 Azure Data Fundamentals exam.

Azure provides multiple storage services for **non-relational data**. You can use these services to store **structured**, **semi-structured**, or **unstructured** data, for use with **databases**, and for providing disks and files to storage resources such as **virtual machines** or **applications**.

In this chapter, you will cover three key areas:

- Describe Azure Blob storage
- Describe Azure Files storage
- Describe Azure Table storage

The topics are related to the *DP-900 Azure Data Fundamentals* certification exam, covering skills for *Describe Considerations for Working with Non-Relational Data on Azure*; this corresponds to about 15-20% of the total measured skills.

By the end of this chapter, you will be ready for the *DP-900 exam*. You will also have learned the skills to implement Azure Storage solutions for real-world applications, enabling you to manage and secure data in the cloud more effectively.

Let's look at the capabilities of storage services within Azure environments.

Azure Storage Services

Azure storage services offer many solutions to meet different storage requirements, forming a crucial part of Microsoft's cloud infrastructure. They cater to varying data and use cases, ensuring scalable, dependable, and affordable options for all kinds of storage needs.

Azure Storage offers three core storage services:

- **Azure Blob service**: Provides a data store for **unstructured data**, accessed via an **API**, used for backup, analytics, web content, images, video, audio, and documents.

- **Azure Files service**: Provides **managed file shares** accessible through **standard file-level protocols** such as **Server Message Block (SMB)** and **Network File System (NFS)**. **New Technology File System (NTFS)** permissions can be set at the folder and file level for access control.

- **Azure Table service**: Provides a **NoSQL key-value pair store** for large amounts of **structured data** with **low latency**.

Figure 4.1 visually represents the three core Azure storage services you will learn about for the exam objectives in this chapter – Azure Blob storage, Azure Files storage, and Azure Table storage:

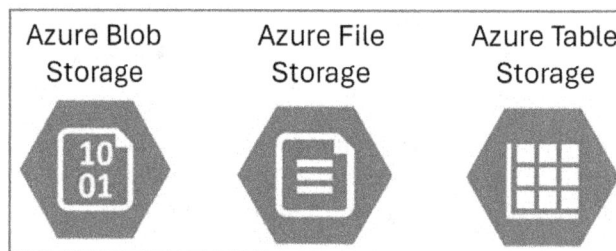

Figure 4.1: Azure storage services for exam objectives

You must understand these three Azure storage services outlined in the *Describe Considerations for Working with Non-Relational Data* section for the *Skills Measured area* of the *DP-900 exam*.

> **Note**
>
> The three storage services shown in *Figure 4.1* are not the only ones available on Azure; for example, there are the **Azure Queue Storage** and **Azure NetApp Files** services; however, for the *DP-900 exam*, skills on these other storage services are not measured. For completeness, though, should you wish to extend your learning about the Azure storage services beyond the exam skills measured areas, you can learn more at `https://learn.microsoft.com/en-us/azure/storage/`.

Next, before you learn about these three storage services, you must understand the concept of storage accounts.

Storage Accounts

Azure storage accounts are the **logical stores** for **objects**. They hold different data types, such as **blobs**, **files**, and **tables**.

You can store all of these storage service types in a **single storage account**, or you can create each service in a **separate storage account**.

You may want each storage service to use its own storage account, as storage accounts offer different redundancy options and monitoring (logging) features. You can independently apply settings such as data tiering, redundancy, encryption, and access controls for each storage service you wish to create.

For **access control list** (**ACL**) support (allowing permissions to be set at both the file and folder levels within the storage account), the storage account must be enabled for a **hierarchical namespace**, which provides the file and folder-level **role-based access control** (**RBAC**) (RBAC is a security mechanism that restricts system access based on the roles of individual users within an organization). This level of access control for data objects within a storage account is provided by the **Azure Data Lake Storage** (**ADLS**) storage solution. For reference, ADLS is a storage service that combines high-performance capabilities with the hierarchical namespace to organize and manage data efficiently, supporting advanced analytics and machine learning workloads.

Storage accounts integrate with other Azure services, making them ideal for data-driven workloads and applications.

Now that you have completed the introduction section to Azure Storage, the following section will teach you about the Azure Blob storage service.

Describe Azure Blob Storage

Binary large objects (**Blobs**) provide extremely scalable and inexpensive object storage and are one of the core building block elements of the Microsoft Azure cloud computing platform.

Blob storage stores unstructured data, such as images and audio files held in logical stores for data that can be retrieved via **API calls** using a globally unique identifier for each stored blob. Workloads "call" the Azure Blob storage API for read and write operations. Each blob item holds **metadata** and has a **unique identifier**.

Blob storage should be used when storing data that does not fit into a file system with traditional file-level access protocols such as SMB and NFS or relational databases such as Microsoft SQL.

Figure 4.2 illustrates data object organization for the Blob storage service:

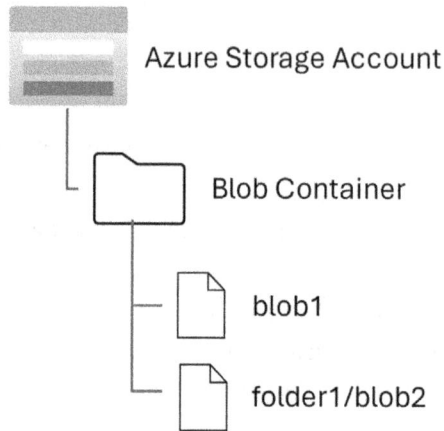

Figure 4.2: Blob storage service data organization

Figure 4.2 shows that Blob storage has a **flat namespace** that uses **virtual directories** for paths to blobs within blob **containers**. A flat namespace in Blob storage means that all data is stored in a single level without hierarchical folders or subdirectories, where each blob is identified by a unique name within the storage container.

An actual data and folder hierarchy and operations at the folder level, such as implementing access controls, require using the ADLS service. **Retention policies** can be set to delete old blob versions automatically, and blobs can be tiered automatically; these are functionalities provided by blob **lifecycle management**. **Versioning** can also be implemented for blob history.

The Azure Blob storage service supports the following blob types:

- **Block blobs**: These are used to store ordered binary files, such as images, documents, audio/media files, and backups.

- **Page blobs**: These are used to store random access files – objects that are randomly written and read in no order, such as databases and **virtual machine disks**.

- **Append blobs**: These are used to store **log information**; objects are consecutively added to the last blob, such as logs, and optimized for sequential writes.

- **ADLS**: This is used to store data for analytics and provides a hierarchical folder structure that allows folder-level operations such as access control.

These data objects make it possible to handle large amounts of unstructured data.

ADLS

ADLS is a massively scalable and innovative cloud storage platform built for **big data analytics** workloads.

ADLS requires creating a storage account with a hierarchical namespace that must be enabled on the storage account level; you cannot revert an ADLS-upgraded storage account to a flat namespace.

Figure 4.3 illustrates data organization for ADLS:

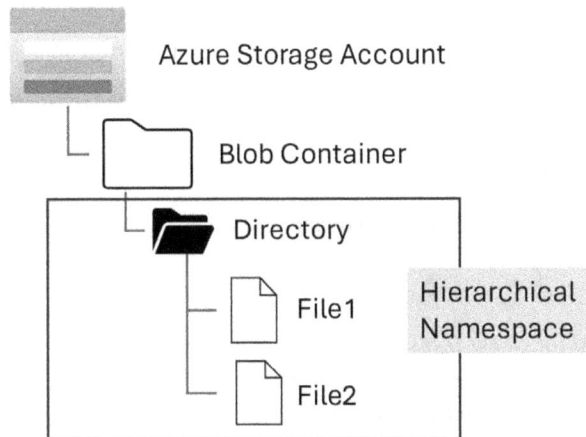

Figure 4.3: ADLS storage service data organization

ADLS provides an ideal platform for storing and analyzing extensive amounts of structured and unstructured data of any size and format. It combines the benefits of a **Hadoop distributed file system** with Azure Blob storage's security, management, and performance capabilities.

The advanced analytics services offered by Azure, such as **Azure HDInsight**, **Azure Databricks**, and **Azure Synapse Analytics**, require a **distributed file system** to be mounted that can be provided by integration with ADLS.

Consider the costs of transferring data across different locations to ADLS.

Storage Tiers

The Azure Blob storage service provides the following tiers:

- **Hot**: This tier offers the **highest cost** with the **lowest latency**
- **Cool**: This tier offers a **lower cost** with **higher latency**
- **Archive**: This tier provides the **lowest cost** with the **highest latency**

You can move your data between these tiers based on changing access patterns and retention needs. Next, you will discover more about each of these tiers.

Hot Tier

The **hot tier** is optimized for frequently accessed data and is suitable for data that requires low-latency access times and high performance; storage media that is of high performance is used to ensure this.

This is the default tier and offers the fastest access to data, with response times in the millisecond range, which is ideal for workloads where fast data access is critical, such as streaming, analytics, or online transactions.

Cool Tier

The **cool tier** is for less actively accessed data, and with slightly longer access times than the hot tier, it offers a lower-cost alternative to store data than the hot tier, as the data is not likely to be accessed for at least 30 days or longer.

The cool tier is ideal for **backup**, **disaster recovery**, and other data archive scenarios.

Archive Tier

The **archive tier** is used for rarely accessed data with **long-term retention** requirements. This tier is designed for data that may not be accessed for at least 180 days. The archive tier provides the lowest storage costs but has the most extended access times, measured in hours. Moving data between an archive tier and a cool tier is a process referred to as **rehydration**.

The archive tier is intended for long-term data archives for compliance, regulatory, or historically enduring purposes that might require storage for years.

You have now learned about the Azure Blob storage service, and next, you will learn about the Azure Files storage service.

Describe Azure Files Storage

Azure Files storage is a file-sharing solution that is fully managed and enables you to share files across multiple virtual machines, applications, and cloud services. Azure Files storage is designed to store and share files such as documents, images, and application data; it uses a storage account for its data and management planes.

It is built on the industry-standard **SMB 3.0** protocol and **Network File System** (**NFS**), meaning you can **mount file shares** (shared folders) so they appear as **mapped drives** and access them as local resources. Mapped drives are networked storage locations that appear as local drives on a computer, allowing users to access and manage files on remote servers as if they were on their own system.

Figure 4.4 illustrates data organization for the Azure Files Storage service:

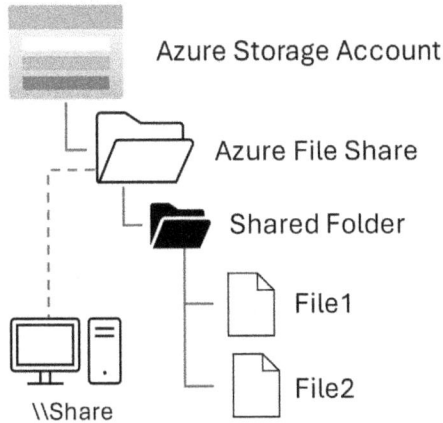

Figure 4.4: Azure Files storage service data organization

Azure Files storage provides the following tiers:

- **Standard (HDD)**: This utilizes **hard disks**
- **Premium (SSD)**: This utilizes **solid-state disks**

Tiers help you choose the right solution for your performance or cost optimization needs.

Different features in Azure Files storage allow the management of file shares easily. For example, you can use the **AZCopy** command line for file transfer operations and the **Azure File Sync** service to synchronize a copy of local data with that stored in Azure Files.

Entra ID (formerly Azure Active Directory) can control access based on identity, allowing you to manage access at the user or group level. You can also use **shared access signatures** to access your file shares temporarily.

Azure Files storage also supports **incremental backups**, allowing you to back up only the changes to your file shares. **Azure Private Link** also enables you to access your file shares securely over a **private endpoint**.

In this section, you gained knowledge of the Azure Files storage service; now, you will explore the Azure Table storage service.

Describe Azure Table Storage

Azure Table storage is a cloud-based NoSQL key-value store with high scalability, flexibility, and performance for storing large amounts of **non-relational structured** data. It is a good fit for workloads that require fast, scalable, and highly flexible data storage.

Figure 4.5 illustrates data organization for the Azure Table service:

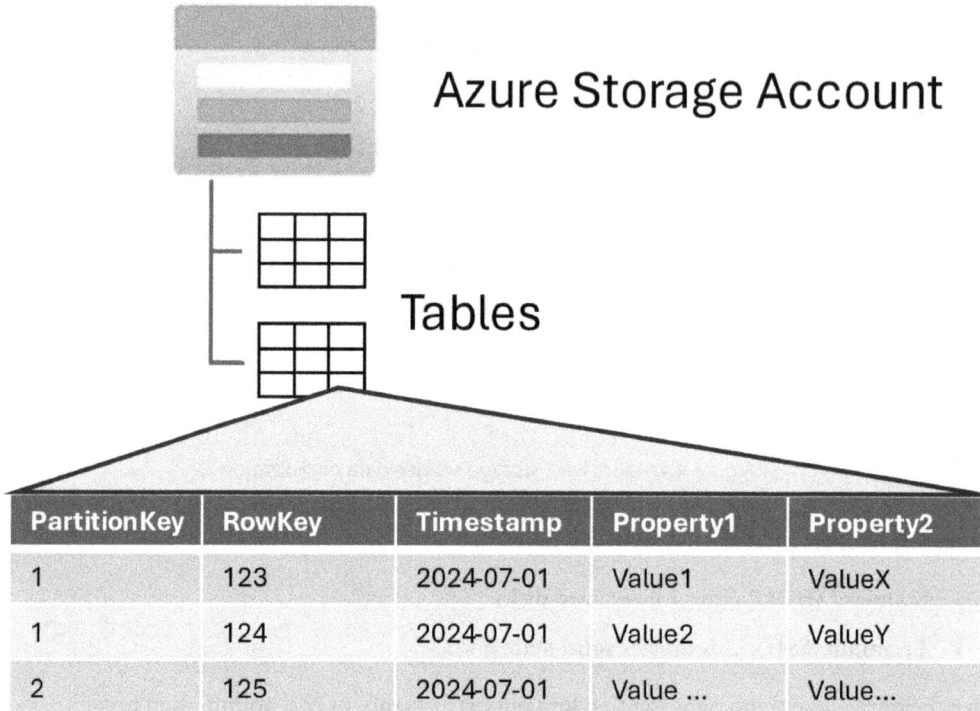

PartitionKey	RowKey	Timestamp	Property1	Property2
1	123	2024-07-01	Value1	ValueX
1	124	2024-07-01	Value2	ValueY
2	125	2024-07-01	Value ...	Value...

Figure 4.5: Azure Table storage service data organization

Table storage allows you to store and query data using a simple schema-less design, which means that if your application needs to change, you can add or remove columns.

Table storage provides several features that enable you to manage your data effectively and speed up access, such as **partitions** (segments used to organize and distribute data across multiple servers for efficient querying and scaling), batch transactions to perform multiple operations atomically, and point-in-time recovery to recover your data to a specific point in time. Azure Table storage allows for an optional read-only secondary region for high availability.

Table storage also supports **Azure Cosmos DB**, which holds data as **relationships** rather than in **tables**, which allows you to run analytics and real-time log entries and queries on your data using popular APIs such as **SQL**, **MongoDB**, **Cassandra**, and **Gremlin**. You must create a storage account before using the Azure Table service.

This section on Azure Table storage concludes this section for the *DP-900 exam* skills measured area *Describe considerations for working with non-relational data on Azure*. This section also concludes the learning content for this chapter. Now, it's time to summarize what skills you have learned in this chapter.

Summary

The chapter outlined the Azure Storage services, which are key to understanding cloud storage options and preparing for the *DP-900 Azure Data Fundamentals exam*.

You looked at storage accounts, which are pivotal for data storage in Azure, supporting various storage types in one account. They feature multiple redundancy choices, detailed access control with hierarchical namespaces, and smooth compatibility with Azure's other services.

You also learned about Azure Blob storage, which is designed for unstructured data such as pictures, video content, and backups. It provides different blob types tailored to specific requirements and storage levels (hot, cool, and archive) that help balance cost against performance. Capabilities such as automated tiering, data versioning, and setting retention policies refine data governance.

Then you looked at Azure Files, which delivers managed file shares that can be accessed using SMB and NFS, simplifying file sharing among virtual machines and apps. It includes features such as identity-based access with Entra ID, incremental snapshots for backups, and Azure File Sync.

Azure Table storage, a NoSQL key-attribute store, is optimized for scalability and versatility. The absence of schema enables adaptability, while partitions enhance how data is handled.

This chapter provided the requisite knowledge for the DP-900 certification, the practical understanding for applying these principles in real-world scenarios of selecting the appropriate storage solutions, and considerations for handling non-relational data on Azure.

The next chapter will cover the *Capabilities and Features of Azure Cosmos DB* to follow the skills measured sequence in *Microsoft's Study Guide for the DP-900 exam*.

Additional Reading

This section provides links to additional exam information and study references:

- DP-900 - Microsoft Azure Data Fundamentals study guide: `https://learn.microsoft.com/en-us/credentials/certifications/resources/study-guides/dp-900`

- DP-900 - Microsoft Azure Data Fundamentals self-directed learning: `https://learn.microsoft.com/en-gb/training/modules/explore-provision-deploy-non-relational-data-services-azure/`

Exam Readiness Drill – Chapter Review Questions

Apart from a solid understanding of key concepts, being able to think quickly under time pressure is a skill that will help you ace your certification exam. That is why working on these skills early on in your learning journey is key.

Chapter review questions are designed to improve your test-taking skills progressively with each chapter you learn and review your understanding of key concepts in the chapter at the same time. You'll find these at the end of each chapter.

> **How to Access These Materials**
>
> To learn how to access these resources, head over to the chapter titled *Chapter 9, Accessing the Online Resources.*

To open the Chapter Review Questions for this chapter, perform the following steps:

1. Click the link – `https://packt.link/DP900Ch04`.

    Alternatively, you can scan the following **QR code** (*Figure 4.6*):

Figure 4.6: QR code that opens Chapter Review Questions for logged-in users

2. Once you log in, you'll see a page similar to the one shown in *Figure 4.7*:

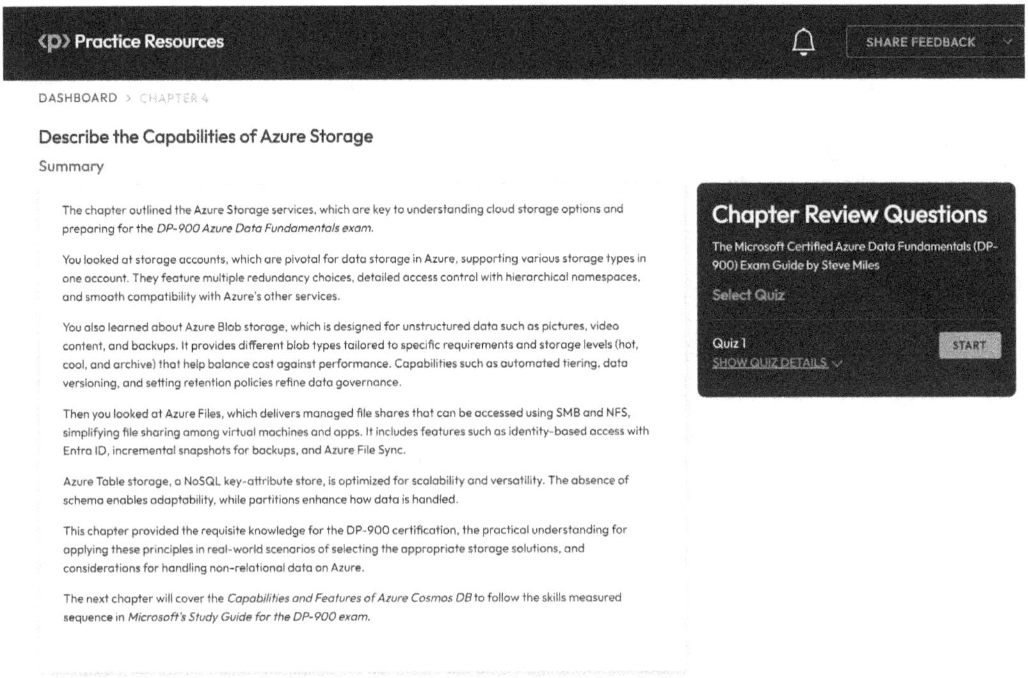

Figure 4.7: Chapter Review Questions for Chapter 4

3. Once ready, start the following practice drills, re-attempting the quiz multiple times.

Exam Readiness Drill

For the first three attempts, don't worry about the time limit.

ATTEMPT 1

The first time, aim for at least **40%**. Look at the answers you got wrong and read the relevant sections in the chapter again to fix your learning gaps.

ATTEMPT 2

The second time, aim for at least **60%**. Look at the answers you got wrong and read the relevant sections in the chapter again to fix any remaining learning gaps.

ATTEMPT 3

The third time, aim for at least **75%**. Once you score 75% or more, you start working on your timing.

> **Tip**
> You may take more than **three** attempts to reach 75%. That's okay. Just review the relevant sections in the chapter till you get there.

Working On Timing

Target: Your aim is to keep the score the same while trying to answer these questions as quickly as possible. Here's an example of how your next attempts should look like:

Attempt	Score	Time Taken
Attempt 5	77%	21 mins 30 seconds
Attempt 6	78%	18 mins 34 seconds
Attempt 7	76%	14 mins 44 seconds

Table 4.1: Sample timing practice drills on the online platform

> **Note**
> The time limits shown in the above table are just examples. Set your own time limits with each attempt based on the time limit of the quiz on the website.

With each new attempt, your score should stay above **75%** while your "time taken" to complete should "decrease". Repeat as many attempts as you want till you feel confident dealing with the time pressure.

5

Describe the Capabilities and Features of Azure Cosmos DB

Due to the rapid digital transformation of modern enterprises, organizations today are looking for flexible, scalable, multi-model, and purpose-built database solutions that can handle various types of data and workloads.

This chapter explores the capabilities and features of Azure Cosmos DB, highlighting the unique offerings that make it a standout choice for many modern applications. You will read about the following main topics:

- Identify use cases for Azure Cosmos DB
- Describe Azure Cosmos DB APIs

You will begin by identifying the various use cases for Azure Cosmos DB, showcasing how its robust architecture supports various scenarios—from real-time analytics and **internet of things** (**IoT**) applications to e-commerce platforms and social networks. Understanding these use cases will help organizations leverage Azure Cosmos DB's strengths in their specific contexts, ensuring optimal performance and efficiency.

Following this, you will explore the different APIs provided by Azure Cosmos DB, each designed to cater to varying application requirements and developer preferences. Whether looking for a document-oriented model with the MongoDB API, a key-value store with the Table API, or a graph-based solution with the Gremlin API, Azure Cosmos DB offers the versatility to support diverse applications.

By the end of this chapter, you will have a comprehensive understanding of Azure Cosmos DB's capabilities, ideal use cases, and the APIs that enable developers to build scalable and high-performance applications in a rapidly evolving technological landscape.

The topics are related to the *DP-900 Azure Data Fundamentals* certification exam, covering measured skills for the *Describe considerations for working with non-relational data on Azure* area; this corresponds to about 15–20% of the total measured skills. Completing this chapter ensures you have the skills to describe the capabilities and features of Azure Cosmos DB and are ready for the certification exam.

You can look at the Azure Cosmos DB use cases in the following section.

Identify Use Cases for Azure Cosmos DB

Azure Cosmos DB is a non-relational NoSQL database **Platform as a Service** (**PaaS**) service that uses a flexible schema for a document-based data store.

It allows you to build highly elastic and available data-connected applications. It offers the ability to quickly access data from anywhere in the world with very low latency and to create it instantly wherever needed. Cosmos DB provides automated indexing, which significantly improves query performance by ensuring that all data is indexed by default, allowing faster access without manual configuration. It also simplifies database management and allows for greater flexibility in data querying.

Features and Capabilities

When using Cosmos DB, you can use its built-in multi-master geo-replication, ensuring your applications are highly available. Geo-replication in Azure Cosmos DB means your data will be automatically replicated across multiple geographic regions, increasing availability and resilience against regional outages. During a regional outage, your application will remain accessible to your users. These data synchronizations will occur in real time, giving users worldwide access to low-latency data and configuring consistency levels per their preference. As such, geo-replication makes Cosmos DB an appealing option for apps with a global footprint, where data availability and performance are two of the key considerations.

Some of the Cosmos DB data models are as follows:

- Key-value (table)
- Column family
- Documents
- Graph

By "Cosmos DB data models," we mean the different data models (forms) the data can take in Azure Cosmos DB (structure, storage, and query). Cosmos DB is a "multi-model" database service, meaning it has data models to suit the different needs of different applications.

Given the number of models, you can see that it is a very flexible solution for many data-connected applications. Cosmos DB allows switching data models, and there is no requirement to change application code. Moreover, Cosmos DB offers automatic indexing that improves query performance; hence, manual indexing is unnecessary.

When creating an Azure Cosmos DB account, you set the API required for the account (and not for the database). Each account can only have one API set (i.e., each account must be created to use a separate API); accounts cannot use multiple APIs. Examples are SQL API, MongoDB API, and Gremlin API; you will learn more about these in the *Describe Azure Cosmos DB APIs* section in this chapter.

As mentioned, you cannot have different APIs in one Azure Cosmos DB account. The API is configured per account and not per database in the account; each database will inherit the API from the account. If you want the database to have different APIs, you will need to create a new account for that database and set the required API at the newly created account for that database (i.e., if you have three databases and each requires a different API, you will need three accounts, with one database in each account).

In Azure Cosmos DB, throughput is set at the container level; each container has its own dedicated throughput provisioned independently of other containers in the same database account.

An Azure Cosmos DB **container** is where data is stored. To begin using Azure Cosmos DB, create an Azure Cosmos DB account in an Azure resource group in your subscription and then create databases and containers within the account.

After learning about Azure Cosmos DB's capabilities in this section, you will discover some use case scenarios in the next section.

Azure Cosmos DB is a globally distributed database service designed to accommodate the demands of next-generation applications with virtually limitless throughput and sub-millisecond latencies across multiple geographies. With global distribution, Azure Cosmos DB can be used in many highly demanding cases for modern applications across many industries. The following are some example usage scenarios for Cosmos DB:

- **Global web and mobile applications**: With its global distribution functionality, Azure Cosmos DB is best suited for web and mobile applications with millions of users who access their data from everywhere, ensuring the best latency and a better user experience.

- **Healthcare applications:** Healthcare organizations can utilize Azure Cosmos DB to store and manage patient records, treatment plans, and clinical data. For instance, a telemedicine platform could store patient data in JSON format, allowing healthcare providers to access and update records in real time. The secure and compliant nature of Cosmos DB also ensures that sensitive health information is managed appropriately while allowing for fast access to critical data during consultations.

- **Gaming**: With its sub-millisecond latency, providing an optimal experience for players worldwide, Azure Cosmos DB can be employed to manage leaderboards for competitive gaming applications. For example, a mobile game could use Cosmos DB to store player scores, achievements, and game statistics in a key-value format for quick lookups. The real-time updates enable players to see their standings instantly, enhancing competition and engagement within the gaming community.

- **Retail and e-commerce**: E-commerce platforms can leverage Azure Cosmos DB to manage product catalogs, customer profiles, and transaction histories. For example, a global retail company can use the document model to store product information in JSON format, including pricing, descriptions, and inventory levels. With automated indexing and low-latency access, customers can quickly search and filter products, enhancing the shopping experience. Additionally, the global distribution features of Cosmos DB ensure that users from different regions experience fast load times.

- **IoT and telemetry data ingestion**: Azure Cosmos DB is well suited for IoT applications, where large volumes of sensor data are generated continuously. For instance, a smart city initiative can utilize Cosmos DB to collect and analyze data from traffic sensors, weather stations, and public transportation systems. The ability to handle massive amounts of ingested data with low latency allows city planners to make data-driven decisions to improve urban infrastructure and optimize traffic flows in real time.

- **Time series data and logging**: With the dynamic schema and elastically scalable architecture, Azure Cosmos DB can be used for time-series data, such as logs, sensor data, monitoring metrics, and more, to enable real-time responsiveness and diagnostics.

Now that you understand some real-world use cases, you will explore the supported APIs.

Describe Azure Cosmos DB APIs

Azure Cosmos DB provides multiple data interaction APIs, such as the API for NoSQL, which is native to Azure Cosmos DB. These APIs can be used when the data is key-value pairs, documents (JSON, XML), columns, or graphs.

The following APIs are supported:

- **SQL API**: Based on a document data store
- **MongoDB API**: Based on a document data store
- **Cassandra API**: Based on a column data store
- **Azure Table API**: Based on a key-value data store
- **Gremlin API**: Based on a graph data store

Each API has use cases that best meet specific needs and preferences. The core SQL API is best for applications needing rich SQL-like querying capabilities, such as e-commerce platforms; the MongoDB API is ideal for document-oriented NoSQL databases, and is often used in social media apps; the Cassandra API is suited for high-throughput key-value stores, which is great for time-series data; the Gremlin API is perfect for graph-based data, such as social networks or fraud detection; and the Table API is a simple key-value store, useful for storing logs or configuration data.

The following sections will explore each of these APIs.

SQL API

The **Cosmos DB SQL API** is a fully managed PaaS NoSQL database service provided by Azure Cosmos DB. Its document-based format supports SQL syntax (such as `SELECT` statements) for querying the stored data and supports JSON data structures. This allows interaction with Cosmos DB using SQL syntax to query, insert, update, and delete data.

The native integration of the NoSQL API within Azure Cosmos DB ensures that you get all the benefits of Cosmos DB, such as automatic scaling, global distribution, low latency, and strong consistency models while working with a NoSQL database paradigm.

This native support for the NoSQL API makes Azure Cosmos DB a versatile and powerful choice for developers looking to build scalable, high-performance applications that require a NoSQL database backend.

MongoDB API

The **MongoDB API** is a fully managed PaaS NoSQL database that supports storing data as **binary JSON (BSON)**, allowing for more efficient encoding and additional data types than JSON. BSON can be faster for certain operations due to its binary format and more space-efficient in some cases, though the efficiency can vary depending on the data structure

The MongoDB API provided by Azure Cosmos DB provides access through a MongoDB-compatible interface to Cosmos DB. Users can interact with Azure Cosmos DB using the same tools and applications as MongoDB, which has a set of official drivers and clients for multiple languages, such as **Language-Integrated Query (LINQ)**, to query MongoDB. Through the MongoDB API, you can use several popular programming languages (Java, Python, Node.js, and others) to migrate your MongoDB workloads to Azure Cosmos DB with little effort.

Cassandra API

Cassandra API is a fully managed database PaaS service based on the **Apache Cassandra API**.

This column-based API gives you a Cassandra-compatible interface to Azure Cosmos DB, allowing you to use Java applications and Apache Cassandra tools that you can use against Azure Cosmos DB. For client apps, the Azure Cosmos DB Cassandra API provides the same easy, familiar Cassandra data-modeling experience as Azure. It includes support for querying items using the **Cassandra Query Language (CQL)**.

Benefits include automatic indexing, built-in sub-millisecond latency guarantees, and global distribution capabilities, making it an excellent fit for powering mission-critical applications.

Table API

The **Table API** is a fully managed NoSQL database PaaS service in Azure. This API is based on a key-value format and is suitable for large-volume transaction applications where data retrieval performance becomes vital; it provides high throughput, low latency, and predictable performance. An example of a transaction that would benefit from using the Table API in Azure could be storing and retrieving user session data for a large-scale online gaming platform.

Each user's session information, such as login time, game state, and progress, is stored as a key-value pair in this scenario. The Table API enables quick access to this data, ensuring low-latency reads and writes, which is crucial for providing a seamless gaming experience where performance and speed are critical.

This API brings an Azure Table storage-compatible interface to Cosmos DB, allowing existing Azure Table storage clients to access Cosmos DB and retrieve key-value pairs. Using the Table API, you can create tables to store your data and then provide a set of rich RESTful APIs to perform a full range of **create, read, update, and delete (CRUD)** operations over your data.

For high availability, data can be replicated across multiple regions; support is provided for multiple read regions and multiple write regions, and writes being accepted in multiple regions reduces write latency for applications that are distributed geographically.

A single-digit milliseconds write latency can be provided anywhere in the world, with an availability **service-level agreement (SLA)** (which is the amount of expected "uptime" of service) of 99.999% for multi-region accounts databases and 99.99% for single regions.

There are additional features of auto-indexing for tables, enabling batch operation scenarios to perform operations over your data and supporting OData querying for data to speed up the development process.

Gremlin API

This API provides a graph-based interface to Cosmos DB; it is a good tool for working with graph data in Azure Cosmos DB.

In a graph database, relationships represent connections or associations between **nodes**. These nodes can represent a wide range of data types, such as people, places, things, or concepts, and they can be connected to other nodes through **edges** to represent relationships between them; edges specify relationships between nodes.

These relationships define how nodes are related to each other and can have properties associated with them, which describe the attributes of the entity they represent. Nodes and edges have properties that provide information about that node or edge, similar to columns in a table. Edges can also have a direction indicating the nature of the relationship.

The **Gremlin API** allows you to interact with nodes and edges in your graph database to query the data. **Gremlin** is the language used primarily to query and traverse a graph database; a Gremlin client library or console would typically be used.

With the Gremlin API and query language, you can perform various graph operations, including adding and removing nodes and edges, updating properties, and executing complex graph traversals.

This section described the Azure Cosmos DB APIs and concluded the learning content for this chapter. The following section summarizes the chapter for you and introduces you to what you will learn in the next chapter of this book.

Summary

In this chapter, you learned about the capabilities and features of Azure Cosmos DB, a flexible, scalable, and multi-model database solution designed for modern enterprises undergoing rapid digital transformation.

This chapter included complete coverage of the *DP-900 Azure Data Fundamentals* exam *Skills Measured* area: *Describe considerations for working with non-relational data on Azure.*

You learned that Azure Cosmos DB supports various scenarios, including real-time analytics, IoT applications, e-commerce platforms, and social networks. Understanding these use cases helps organizations leverage Azure Cosmos DB's strengths for optimal performance and efficiency.

Next, you learned that Azure Cosmos DB offers various APIs to cater to different application requirements and developer preferences. These include the SQL API, MongoDB API, Cassandra API, Azure Table API, and Gremlin API. Each API supports specific data interaction paradigms, such as key-value, documents, columns, or graphs.

At the end of this chapter, you have gained a comprehensive understanding of Azure Cosmos DB's capabilities, ideal use cases, and the APIs that enable developers to build scalable and high-performance applications in a rapidly evolving technological landscape.

The next chapter will include coverage of the topic of the *Describe Common Elements of Large-scale Analytics* area in Microsoft's study guide for the *DP-900* exam.

Additional Reading

This section provides links to additional exam information and study references:

- DP-900 – Microsoft Azure Data Fundamentals study guide: `https://learn.microsoft.com/en-us/credentials/certifications/resources/study-guides/dp-900`

- DP-900 – Microsoft Azure Data Fundamentals self-directed learning:

- `https://learn.microsoft.com/en-gb/training/modules/explore-non-relational-data-stores-azure/`

Exam Readiness Drill – Chapter Review Questions

Apart from a solid understanding of key concepts, being able to think quickly under time pressure is a skill that will help you ace your certification exam. That is why working on these skills early on in your learning journey is key.

Chapter review questions are designed to improve your test-taking skills progressively with each chapter you learn and review your understanding of key concepts in the chapter at the same time. You'll find these at the end of each chapter.

> **How to Access These Materials**
>
> To learn how to access these resources, head over to the chapter titled *Chapter 9, Accessing the Online Resources*.

To open the Chapter Review Questions for this chapter, perform the following steps:

1. Click the link – `https://packt.link/DP900Ch05`.

    Alternatively, you can scan the following **QR code** (*Figure 5.1*):

Figure 5.1 – QR code that opens Chapter Review Questions for logged-in users

2. Once you log in, you'll see a page similar to the one shown in *Figure 5.2*:

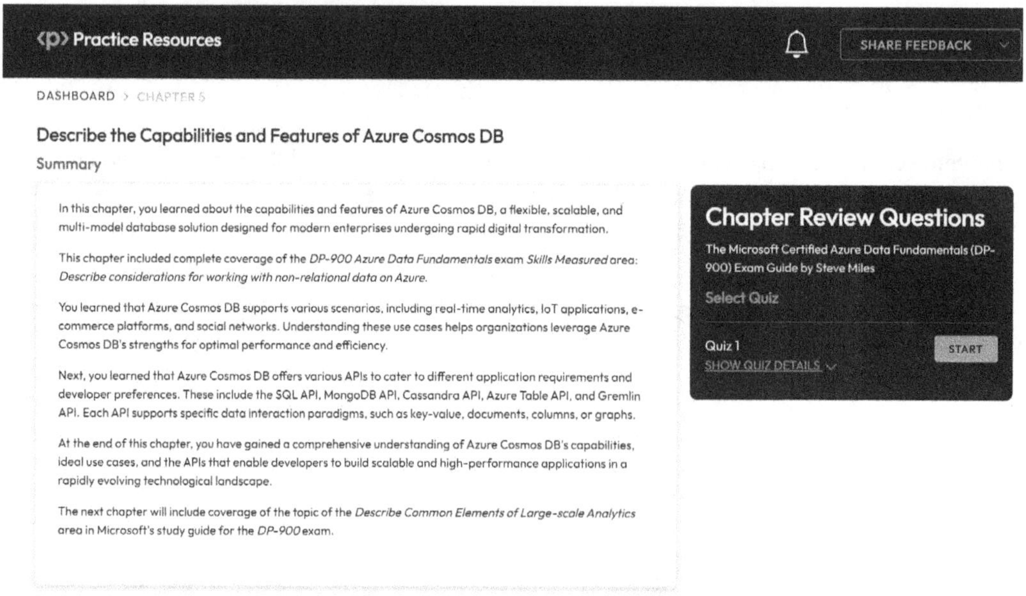

<p> Practice Resources   🔔 SHARE FEEDBACK ⌄

DASHBOARD > CHAPTER 5

Describe the Capabilities and Features of Azure Cosmos DB
Summary

In this chapter, you learned about the capabilities and features of Azure Cosmos DB, a flexible, scalable, and multi-model database solution designed for modern enterprises undergoing rapid digital transformation.

This chapter included complete coverage of the *DP-900 Azure Data Fundamentals* exam *Skills Measured* area: *Describe considerations for working with non-relational data on Azure.*

You learned that Azure Cosmos DB supports various scenarios, including real-time analytics, IoT applications, e-commerce platforms, and social networks. Understanding these use cases helps organizations leverage Azure Cosmos DB's strengths for optimal performance and efficiency.

Next, you learned that Azure Cosmos DB offers various APIs to cater to different application requirements and developer preferences. These include the SQL API, MongoDB API, Cassandra API, Azure Table API, and Gremlin API. Each API supports specific data interaction paradigms, such as key-value, documents, columns, or graphs.

At the end of this chapter, you have gained a comprehensive understanding of Azure Cosmos DB's capabilities, ideal use cases, and the APIs that enable developers to build scalable and high-performance applications in a rapidly evolving technological landscape.

The next chapter will include coverage of the topic of the *Describe Common Elements of Large-scale Analytics* area in Microsoft's study guide for the *DP-900* exam.

Chapter Review Questions

The Microsoft Certified Azure Data Fundamentals (DP-900) Exam Guide by Steve Miles

Select Quiz

Quiz 1   START
SHOW QUIZ DETAILS ⌄

Figure 5.2 – Chapter Review Questions for Chapter 5

3. Once ready, start the following practice drills, re-attempting the quiz multiple times.

Exam Readiness Drill

For the first three attempts, don't worry about the time limit.

ATTEMPT 1

The first time, aim for at least **40%**. Look at the answers you got wrong and read the relevant sections in the chapter again to fix your learning gaps.

ATTEMPT 2

The second time, aim for at least **60%**. Look at the answers you got wrong and read the relevant sections in the chapter again to fix any remaining learning gaps.

ATTEMPT 3

The third time, aim for at least **75%**. Once you score 75% or more, you start working on your timing.

> **Tip**
> You may take more than **three** attempts to reach 75%. That's okay. Just review the relevant sections in the chapter till you get there.

Working On Timing

Target: Your aim is to keep the score the same while trying to answer these questions as quickly as possible. Here's an example of how your next attempts should look like:

Attempt	Score	Time Taken
Attempt 5	77%	21 mins 30 seconds
Attempt 6	78%	18 mins 34 seconds
Attempt 7	76%	14 mins 44 seconds

Table 5.1 – Sample timing practice drills on the online platform

> **Note**
> The time limits shown in the above table are just examples. Set your own time limits with each attempt based on the time limit of the quiz on the website.

With each new attempt, your score should stay above **75%** while your "time taken" to complete should "decrease". Repeat as many attempts as you want till you feel confident dealing with the time pressure.

6

Describe the Common Elements of Large-Scale Analytics

In this chapter, you will gain valuable insights into the fundamental components of large-scale analytics, equipping you with the necessary skills to effectively manage and analyze data in today's data-driven environment. The primary skill being taught here is data analytics architecture, which encompasses designing and implementing systems that enable efficient data ingestion, processing, and storage for analytical purposes.

You will learn about the following main exam questions:

- Describe considerations for data ingestion and processing
- Describe options for analytical data stores
- Describe Azure services for data warehousing, including Azure Synapse Analytics, Azure Databricks, Microsoft Fabric, Azure HDInsight, and Azure Data Factory

The topics are related to the *DP-900 Azure Data Fundamentals* certification exam, covering measured skills for *Describe an analytics workload on Azure*; this corresponds to about 25–30% of the total measured skills.

By the end of this chapter, you will have a comprehensive understanding of large-scale analytics architecture, equipping you with the knowledge to make informed decisions about data ingestion, storage options, and using Azure services in real-world scenarios. This skill set is increasingly important as organizations strive to leverage data for competitive advantage in an ever-evolving marketplace.

Completing this chapter ensures that you have the necessary skills ready for the certification exam.

We will start this chapter by looking at some data ingestion and processing considerations.

Describe Considerations for Data Ingestion and Processing

Depending on an organization's specific needs and requirements, the technologies used to provide data analytics solutions and data ingestion and processing needs can vary.

When considering data ingestion and processing, organizations must evaluate several factors to select technologies that align with their needs. Key considerations include the variety of data types (structured, semi-structured, and unstructured), the volume of data being processed, the preference for real-time versus batch processing, integration with existing systems, scalability, performance, and cost constraints.

For example, Netflix utilizes real-time data ingestion and warehousing to analyze user behavior and optimize content recommendations efficiently. Airbnb adopts a data lake architecture for storage and batch processing, which helps them analyze large datasets related to user preferences and pricing strategies. Similarly, Uber uses real-time data streaming to enhance ride dispatching and user experiences.

However, data analytics solutions generally include common architectural elements – data sources (which can be streamed or non-streamed), data storage (which can be data warehouses or lakehouses), data processing, consumption, and visualization.

Figure 6.1 shows these data analytics architecture elements:

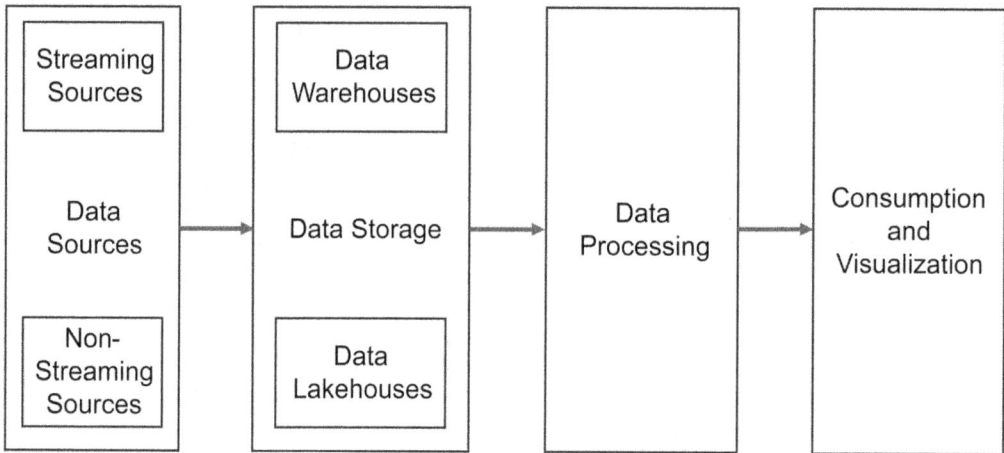

Figure 6.1: The key elements of data ingestion and processing

These key elements of a large-scale analytics solution are explored further in the upcoming sections.

Data Sources

Large-scale analytics begins with data collection and ingestion from various sources, including transactional databases, social media, IoT devices, and web logs. Technologies such as Apache Kafka and cloud services such as Azure Data Factory are commonly used to ingest data efficiently from multiple sources into a centralized analytics-optimized repository.

Data Storage

For large amounts of data to be analyzed, you will need scalable and cost-effective storage for storage solutions such as data lakes (Azure Data Lake Storage), data warehouses, or data lakehouses (Azure Synapse Analytics).

Well-structured data management principles such as partitioning, sharding, compression, and indexing are essential for making data performant and accessible.

Data Processing

Analytics tools such as **Azure Synapse Analytics**, as well as cloud machine learning services such as Azure Machine Learning, provide the means to analyze and learn from data at scale using advanced techniques such as predictive analytics, **natural language processing** (**NLP**), and deep learning to enrich the data and then provide meaningful insights.

Cloud-native data integration processes and **extract, transform, load** (**ETL**)/**extract, load, transform** (**ELT**) services, such as **Azure Data Factory**, are core components of **Microsoft Fabric** and can help automate and streamline data workflows. When you must perform large-scale data ingestion, you create data pipelines to orchestrate ETL/ELT processes.

ETL/ELT processes are different ways of transforming data for analysis. In ETL services, data is extracted from sources, transformed for analysis, and loaded into a target system. In ELT services, data is extracted and loaded into a target system before transformation.

You can create pipelines and run them using Azure Data Factory or a similar pipeline engine in Azure Synapse Analytics or Microsoft Fabric, orchestrating all the components of your data analytics solution in one single-pane-of-glass experience. In the context of data analytics solutions such as Azure Data Factory, Azure Synapse Analytics, or Microsoft Fabric, this means you can orchestrate and oversee all components – data ingestion, transformation, and visualization – from one central interface. This enhances efficiency, improves decision-making, and provides a more user-friendly experience by reducing complexity and streamlining workflows.

The orchestration element utilizes pipelines comprising one or more activities and associated data. An input dataset contains the data that feeds the activities, and an activity consumes an input and produces an output. It can be thought of as a data flow that incrementally transforms the data given to it until it produces an output dataset. Pipelines can be connected with external data sources that allow for the integration of diverse data services.

Ensuring seamless data integration from disparate sources is critical. Integration elements are also provided in platforms such as Azure Databricks that facilitate these transformations by connecting different data systems and ensuring data consistency. Tools such as Apache Hadoop and Apache Spark batch large datasets. These frameworks allow for the parallel processing of data across distributed computing environments. Stream processing frameworks such as Apache Storm and Spark Streaming are employed to process data in motion for real-time analytics.

For further information or a refresher on the ETL/ELT processes, refer to the *Describe Features of Analytical Workloads* section in *Chapter 1, Describe Core Data Concepts*.

Data Consumption and Visualization

Visual representations of data play an important role in interpreting analytics results. The dashboard and reports on Power BI can be interactive, making it easy for stakeholders and end users to analyze high-level results and insights. Collaboration features, such as sharing dashboards and reports and adding annotations, allow users to collaborate and derive insights collectively.

Integrating data consumption and visualization components requires effective data modeling, which involves defining the structure of data and the relationships between data entities. Data marts, subsets of data warehouses focused on specific business areas, help organize data for specific analytical needs.

Large-scale analytics comprises many integrated elements; it is the process that ingests, processes, analyses, consumes, and visualizes data, utilizing different technologies, tools, and practices to process, store, and analyze it. Understanding and applying these common elements guides organizations to use data analytics to inform decisions, improve operations, and be more competitive.

Describe Options for Analytical Data Stores

As our world and lives become increasingly data-led, organizations constantly search for the best way to store, manage, and analyze large volumes of data.

There are two ways to model data, which have led to the rise of two main data solutions in this category – data warehouses and data lakehouses.

Although both of them are used to store information, they are, in fact, completely different in their structure and their functions.

One requirement for understanding the correct choice and application for these technologies would be differentiating between a data warehouse and a data lakehouse. While these technologies have similar usages, they exhibit some differences in their structure and performance.

Here, you will learn about these two solutions and how the underlying technologies of data warehouses and data lakehouses work. You will also get to compare each solution by examining these use cases and how each solution handles them differently. Which one is better suited to different kinds of data-driven use cases?

Understanding these differences can help you choose the right strategy for your organization's data analytics goals and outcomes.

Data Warehouses

Data warehouses maintain a central repository for storing aggregated data from various source systems in an organized, structured relational format. They are designed to deliver optimal performance and support report-generation activities. They offer excellent query performance and are ideal for executing **business intelligence (BI)** operations, complex queries, and analytical processing.

Figure 6.2 provides a visualization of a data warehouse data architecture.

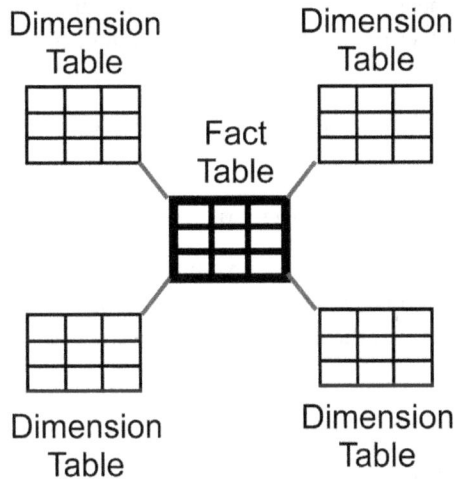

Figure 6.2: A data warehouse data architecture representation

Data warehouses are built on the star schema (*Figure 6.2*). It's called a star schema because the fact table, which contains most of the data, is in the center, connected to smaller dimension tables, like the spokes on a wheel.

In this schema design, a central fact table contains quantitative data for analysis, such as sales revenue or transaction counts. This fact table is surrounded by dimension tables that provide contextual information, such as product details or periods.

This schema is a fundamental and widely used design that simplifies data modeling and enhances query performance because its architecture mimics the structure of a query – cube aggregation, dimension queries, and so on.

The key components of a **star schema** are as follows:

- **Fact table**:

  - **Central repository**: The fact table is the core of the star schema, containing quantitative data and metrics subject to analysis.

  - **Measures**: They hold numeric values, such as sales revenue, quantities sold, or transaction amounts, known as measures or facts.

  - **Foreign keys**: The fact table includes foreign keys that reference the primary keys in the dimension tables, linking them together.

- **Dimension tables**:

  - **Contextual data**: Dimension tables provide descriptive information, or attributes, related to the facts in the table. These might include time, geography, product, or customer dimensions.

  - **Attributes**: Each dimension table has a set of attributes that describes the dimension in more detail – for example, a product dimension table might contain attributes such as product name, category, and brand.

  - **Primary keys**: Each dimension table will have a single primary key that uniquely identifies each row.

A star schema has advantages, such as simplicity, query performance, flexibility, and scalability.

However, due to the nature of denormalization with dimension tables, redundant data will be contained within them, which can lead to increased storage. There can be a certain level of complexity for the ETL process with this schema when the data sources are large and diverse.

You will learn more about schemas in the *Describe Core Concepts of Data Modeling* section in *Chapter 8, Describe Data Visualization in Microsoft BI*.

Some of the key characteristics of data warehouses are as follows:

- **Structured data storage**: Data warehouses store structured data – information that is highly organized, such as financial data that is often stored in tables, with each column corresponding to a predefined schema, such as employee name, employee ID, and department.

- **An ETL process**: This flow extracts data from many sources, transforms it into a standard structure, and loads it into the warehouse. The ETL process ensures data quality and consistency.

- **Optimization for read operations**: Data warehouses are optimized for read-heavy operations ("query load"), making them a good fit for reports and dashboards.

- **High performance**: With indexing, partitioning, and optimized storage techniques, data warehouses deliver fast query performance.

Some use cases of data warehouses are BI and analytics, reporting and dashboarding, historical data analysis, and data mining.

A data warehouse is a database system that stores aggregated structured data for BI and analytics. It has advantages such as high query performance, data consistency, and advanced analytical capabilities, but it also has limitations such as high cost, limited flexibility, and a time-consuming ETL process.

A data lakehouse is a hybrid approach that combines the features of data warehouses and data lakes, allowing for unified storage, schema-on-read, integrated analytics, and cost efficiency.

Data Lakehouses

In a data lakehouse, raw data can be stored in its native formats without any removed attributes. Although a data warehouse is a comparable approach to managing data, a data lakehouse also natively stores structured and raw data at a low cost and can deliver an analytic capability, similar to that of a data warehouse system for real-time analytics.

Figure 6.3 illustrates a data lakehouse architecture. A data lakehouse combines the concepts of data warehouses and data lakes to improve flexibility and scalability. The architecture comprises a distributed filesystem that stores raw data files, with a relational storage layer to expose them as tables.

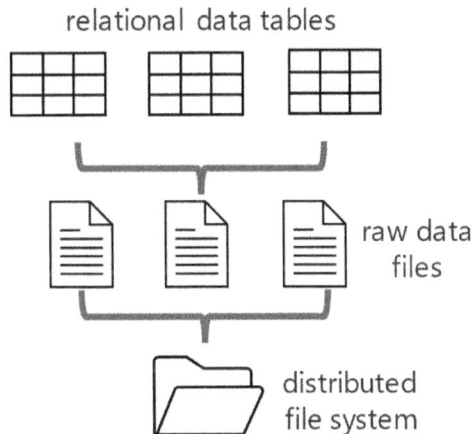

Figure 6.3: A data lakehouse architecture representation

Some of the key characteristics of data lakehouses are as follows:

- **Unified storage**: A single repository can be used for structured, semi-structured, and unstructured data, which is stored together in a data lakehouse

- **Schema-on-read**: Unlike data warehouses, which require predefined schemas, a data lakehouse has a schema-on-read architecture, which means that data can be ingested in various formats without any adapted schema defined upfront, unlike a data warehouse

- **Integrated analytics**: With the self-service capabilities of analytics platforms such as Microsoft Fabric, machine learning, and stream processing integration, data lakehouses allow users to explore, model, and make decisions based on integrated data
- **Cost efficiency**: The scalability based on data lakes ensures that data lakehouses ingest data at a lower cost than data warehouses

Data lakehouses are used in advanced analytics and machine learning, real-time data processing, large-scale data integration, and hybrid data management. The advantages of data lakehouses are that they offer flexibility in handling various data types. They also offer scalability and cost efficiency, enhanced analytical capabilities, and a simplified data architecture. However, data lakehouses are an emerging technology, so best practices are ever-evolving, and there could be potential challenges related to data governance, security, and integration complexity.

In the next section, we will look at the considerations when choosing the right solution for an analytical data store.

Considerations for Data Warehouses

When considering the implementation of a data warehouse, it's crucial to evaluate various factors that can significantly impact its effectiveness and suitability for your organization. The following are some key considerations that can guide your decision-making process. Understanding these factors will help ensure that your data warehouse meets your business requirements and supports your analytics and reporting needs:

- **Data structure**: If your data tends to be highly structured and you are doing lots of high-performance querying and reporting, consider a data warehouse
- **BI**: Data warehouses are a good solution for organizations whose BI is mature and whose reports have business needs for auditability and consistency

Considerations for Data Lakehouses

Assessing the deployment of a data lakehouse requires attention to some important factors that highlight its distinct advantages and appropriateness for an organization. Understanding these elements ensures that you can evaluate whether a data lakehouse is consistent with your overall data strategy and accommodate the varied analytical requirements of your business:

- **Data variety**: A data lakehouse provides maximum flexibility if your organization handles various data formats, from unstructured and semi-structured to structured data
- **Scalability and cost**: A data lakehouse might be the best option if your organization needs to scale storage and simplify advanced queries and real-time analytics

- **Innovation and flexibility**: If you are exploring innovative applications for data, such as machine learning, and are open to a flexible data architecture itself, a data lakehouse can be a great fit

Data lakehouses are the next evolution and innovation path of data management, offering a cohesive environment that is scalable and low-cost for combining different data types and high-performance analytics.

In this section, you learned how to describe analytical data store options. Next, we will explore the data warehousing services on Azure.

Describe Azure Services for Data Warehousing

Microsoft Azure provides the most complete end-to-end scalable data analytics architectures possible in the current technology landscape. The data warehousing functions provided by Azure such as Azure Synapse Analytics, formerly Azure SQL Data Warehouse, are a core component of the Microsoft Fabric **software as a service (SaaS)**-based universal analytics platform and deliver the most powerful integrated architectures available to any company to ingest, process, analyze, and consume enriched data.

Azure's data warehousing architecture is built upon a series of core components that collaborate to create a high-performance, scalable, and secure data solution.

Figure 6.4 provides a high-level visualization of Microsoft analytics solutions; this breaks down those provided to you as a **platform as a service (PaaS)** service and those provided to be consumed as a SaaS platform.

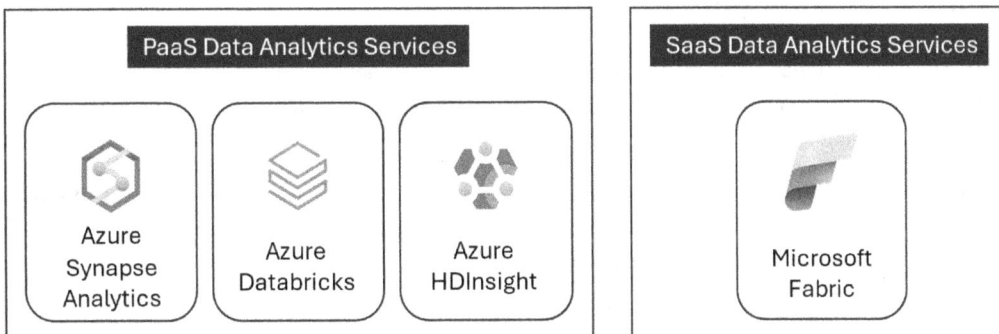

Figure 6.4: Microsoft analytics solutions

The content covered in the following section delves into Azure's data warehousing architecture solutions, as shown in *Figure 6.4*.

Azure Data Factory

Azure Data Factory is a cloud-based data integration service for orchestrating and automating data movements and transformations. ETL and ELT can be leveraged to ingest data from heterogeneous sources into a data analytics data store, such as a data warehouse or lakehouse (for example, Microsoft Fabric or Azure Synapse Analytics).

Azure Data Factory is a valuable tool for businesses that want to integrate and automate their data processes. It enables organizations to move and transform data from various sources into data stores such as data warehouses or lakehouses, making analysis easier and more effective.

In today's data-driven environment, companies generate significant amounts of data from different sources, such as sales systems, online platforms, and inventory management. They need a way to collect and analyze this data to improve decision-making. For example, consider a retail company that operates both physical stores and an online shop. This company collects data from in-store sales transactions, customer interactions from online orders, and inventory management systems. They want to analyze this data to understand customer behavior and optimize inventory management.

Azure Data Factory allows for real-time data updates, enabling the company to monitor sales and inventory levels as they happen. This capability helps the company quickly change marketing efforts or restock popular items. Finally, Azure Data Factory can schedule these processes to ensure that the latest data is always ready for analysis, reducing manual work, ensuring data consistency and integrity, and minimizing errors.

Azure Synapse Analytics

Azure Synapse Analytics is a unified analytics service that combines big data and data warehousing into a **massively parallel processing** (**MPP**) engine and federates the ingestion, preparation, management, and serving of data for BI and **machine learning** (**ML**) purposes.

Microsoft Fabric's key component is at the storage layer with Azure Data Lake Storage. It is intended to scale for exabytes of data and work with Azure Synapse Analytics, also part of Microsoft Fabric and a data lakehouse architecture type of data warehouse.

Azure Synapse Analytics can also be integrated with Azure Machine Learning, which provides an end-to-end data science solution by allowing data scientists to prepare data, create experiments, and deploy models at scale.

Lastly, the business analytics service is Power BI, also part of Microsoft Fabric, which offers interactive visualization and BI capabilities. It can be connected directly to Azure Synapse Analytics for real-time data analysis and reporting.

Azure HDInsight

Azure HDInsight is a fully managed service on Microsoft Azure that lets organizations process big data using open source frameworks, such as Apache Hadoop, Apache Spark, Apache Hive, Apache HBase, Apache Kafka, and Apache Storm. Azure HDInsight helps gather and analyze data extremely fast, flexibly, and cost-effectively, and it shares results with many end users, both inside and outside the company, for key decision-making strategies.

You can see the role of each supported open source framework as follows:

- **Apache Hadoop**: For batch processing
- **Apache Spark**: For in-memory data processing
- **Apache Hive**: For data warehousing
- **Apache HBase**: For NoSQL databases
- **Apache Kafka**: For real-time data streaming
- **Apache Storm**: For real-time event processing

As you can see, Azure HDInsight offers a reliable workload management platform for big data, which agile businesses can utilize to process large amounts of data and gain actionable intelligence.

Azure Databricks

An example of an optimized analytics and ML operating platform, Azure Databricks leverages the performance of an Apache Spark runtime engine that is tuned to specific Azure cloud services. Designed for use by data scientists, data engineers, and business analysts, the environment accelerates the turnaround of complex analytics and ML workloads at scale, easily and collaboratively. Jointly developed alongside Apache Spark's originators, this is a widely used, open source, and unified analytics engine, designed for data integration and processing across various sources and formats.

Azure Databricks allows users to integrate Apache Spark/Spark cluster management seamlessly with a comprehensive and integrated environment and Azure services. It is a scalable resource management solution that can be used with an open source, free version, or with commercial and premium services and features. It's a high-performance environment for big data analytics and ML workloads. Currently, besides Spark (PySpark, Scala, and Java), Azure Databricks understands languages that include Python, SQL, and R. It integrates natively with Azure Data Lake Storage, Azure SQL Data Warehouse, Azure Cosmos DB, Azure Machine Learning, Azure Synapse Analytics, and Power BI. This simplifies data engineering and data science workflows.

The following outlines an example workflow for Azure Databricks:

1. **Data ingestion**: Use Databricks to ingest data from Azure Blob Storage.

2. **Data transformation**: Clean and transform the data using Apache Spark within Databricks.

3. **Model training**: Train an ML model using Spark's MLlib or other supported ML libraries.

4. **Data storage**: Store the processed data and model outputs in Azure Data Lake Storage.

5. **Visualization**: Create interactive dashboards and reports using Power BI, connected to the data processed in Databricks.

In conclusion, Azure Databricks is a unified analytics platform for big data and ML that provides advanced analytics, scalable computing, and a user-friendly interface. This is an important tool for companies looking to leverage their data and improve the quality of their output.

Microsoft Fabric

Microsoft Fabric is a SaaS-based analytics platform that provides real-time data analytics, using multiple continuous data ingestion sources. Microsoft Fabric uses Azure Synapse Analytics as its compute service, while Eventstream is used to persist ingested real-time streaming source data in a lakehouse or KQL database table. The real-time data can be queried using SQL and KQL.

Users can have meaningful insights at their fingertips by exploiting the analytics, data integration, and visualization capabilities of this versatile, all-round, purpose-built unified analytics platform. Its highly intuitive and usable persona-driven design and myriads of tools and features make it a compelling choice to enable well-informed, data-inspired decision-making.

As you can see, Azure provides data professionals with a comprehensive toolset of services and capabilities that enable businesses to derive the maximum value from their data and make smarter decisions, while facilitating innovation. To fully exploit the potential of the Azure architecture for data warehousing, data scientists, engineers, analysts, and business leaders have to understand the Azure data landscapes it provides.

This section concludes the learning content for this chapter. Now, it is time to summarize what skills you have learned in this chapter.

Summary

This chapter included complete coverage of the *DP-900 Azure Data Fundamentals* exam's *Skills Measured area – Describe the Common Elements of Large-scale Analytics.*

You started this chapter by exploring considerations for data ingestion and processing, and understanding how to select appropriate technologies based on data variety, volume, and processing needs. This foundational knowledge is crucial, as effective data ingestion ensures that the right data is captured and made available for analysis promptly.

Then, the chapter delved into options for analytical data stores, discussing different architectures such as data warehouses, data lakes, and data lakehouses. Learning about these options is essential for understanding how to store and manage data in a way that supports diverse analytical workloads and enables actionable insights.

Finally, you explored the various Azure services for data warehousing, including Azure Synapse Analytics, Azure Databricks, Microsoft Fabric, Azure HDInsight, and Azure Data Factory. Each service offers unique features and capabilities that can be leveraged to build scalable and efficient analytics solutions in the cloud. Understanding these services and their applications is vital for organizations seeking to harness the power of data to drive business decisions, enhance operational efficiency, and foster innovation.

The next chapter will describe the consideration of real-time data analytics to follow the skills measured in Microsoft's study guide for the DP-900 exam.

Additional Reading

This section provides links to additional exam information and study references:

- DP-900 – Microsoft Azure Data Fundamentals study guide: `https://learn.microsoft.com/en-us/credentials/certifications/resources/study-guides/dp-900`

- DP-900 – Microsoft Azure Data Fundamentals self-directed learning: `https://learn.microsoft.com/en-gb/training/modules/examine-components-of-modern-data-warehouse/`

Exam Readiness Drill – Chapter Review Questions

Apart from a solid understanding of key concepts, being able to think quickly under time pressure is a skill that will help you ace your certification exam. That is why working on these skills early on in your learning journey is key.

Chapter review questions are designed to improve your test-taking skills progressively with each chapter you learn and review your understanding of key concepts in the chapter at the same time. You'll find these at the end of each chapter.

> **How to Access These Materials**
>
> To learn how to access these resources, head over to the chapter titled *Chapter 9, Accessing the Online Resources.*

To open the Chapter Review Questions for this chapter, perform the following steps:

1. Click the link – `https://packt.link/DP900Ch06`.

    Alternatively, you can scan the following **QR code** (*Figure 6.5*):

Figure 6.5: QR code that opens Chapter Review Questions for logged-in users

2.   Once you log in, you'll see a page similar to the one shown in *Figure 6.6*:

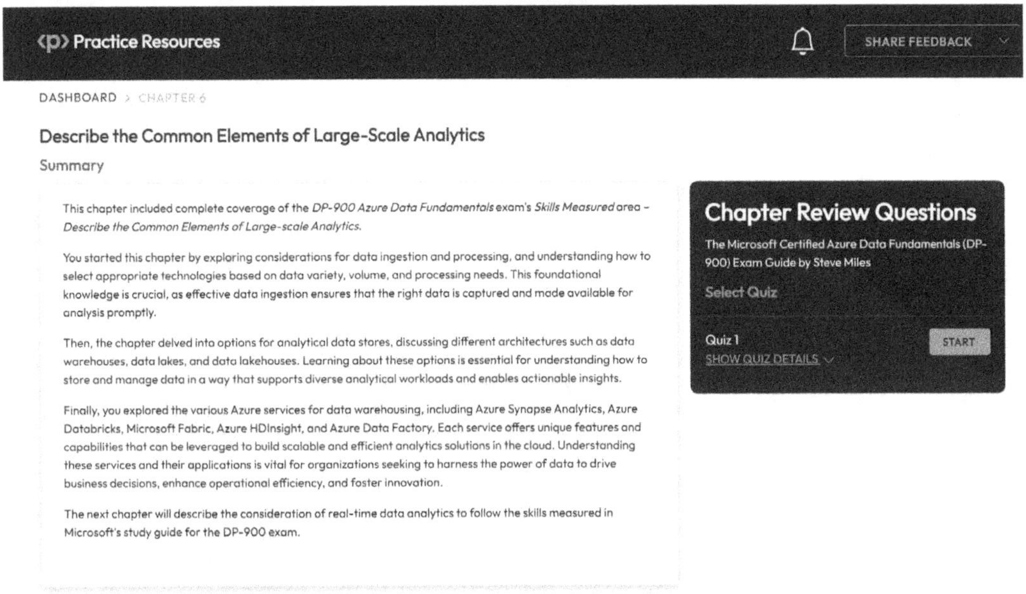

Figure 6.6: Chapter Review Questions for Chapter 6

3.   Once ready, start the following practice drills, re-attempting the quiz multiple times.

Exam Readiness Drill

For the first three attempts, don't worry about the time limit.

ATTEMPT 1

The first time, aim for at least **40%**. Look at the answers you got wrong and read the relevant sections in the chapter again to fix your learning gaps.

ATTEMPT 2

The second time, aim for at least **60%**. Look at the answers you got wrong and read the relevant sections in the chapter again to fix any remaining learning gaps.

ATTEMPT 3

The third time, aim for at least **75%**. Once you score 75% or more, you start working on your timing.

Tip

You may take more than **three** attempts to reach 75%. That's okay. Just review the relevant sections in the chapter till you get there.

Working On Timing

Target: Your aim is to keep the score the same while trying to answer these questions as quickly as possible. Here's an example of how your next attempts should look like:

Attempt	Score	Time Taken
Attempt 5	77%	21 mins 30 seconds
Attempt 6	78%	18 mins 34 seconds
Attempt 7	76%	14 mins 44 seconds

Table 6.1: Sample timing practice drills on the online platform

---

Note

The time limits shown in the above table are just examples. Set your own time limits with each attempt based on the time limit of the quiz on the website.

With each new attempt, your score should stay above **75%** while your "time taken" to complete should "decrease". Repeat as many attempts as you want till you feel confident dealing with the time pressure.

Describe Consideration for Real-Time Data Analytics

Sometimes, **big data analytics** needs to be done in near-real time – for example, in some domains of financial trading, fraud detection, or cybersecurity. For that purpose, Big Data real-time analytical solutions are required.

For instance, with the help of some real-time data analytics solutions, you can easily determine stock price trends and make predictions on the market to make trading decisions. You can also learn about potential frauds committed against your accounts using real-time data analytics solutions; user and transactional data helps a great deal in discovering frauds.

Real-time data analytics solutions train machine learning models on large amounts of historical data from various sources. They can identify patterns, anomalies, and trends faster and more precisely than a human worker ever could. Up-to-the-minute or near-real-time insights, produced by real-time data analytics solutions, are typically exhibited through visualizations, dashboards, and reports shown to decision-makers, who use these visual presentations to make decisions.

Real-time data analytics solutions, such as Apache Kafka, Apache Spark Streaming, Azure Stream Analytics, and others, can read and process multiple terabytes of data per second, analyze it, execute machine learning algorithms, and generate dashboards, visualizations, and reports that can be read and acted upon by decision-makers. Real-time data analytics solutions are also increasingly being integrated with other tools and systems, such as analytical data stores and business intelligence tools, to give an organization the whole picture of its data.

In this chapter, you are going to cover the following main topics:

- Describe the difference between batch and streaming data
- Identify Microsoft cloud services for real-time analytics

In this chapter, you will learn how to identify and describe the key concepts and solutions related to real-time data analytics on the Microsoft Azure platform. You will begin by exploring the fundamental differences between batch and streaming data, two primary methods of processing data that cater to different organizational needs. Following this, you will learn how to describe the various Azure data platform services that facilitate real-time analytics, providing insights into their capabilities and use cases.

This chapter aligns with the learning content for the *DP-900 Azure Data Fundamentals* exam, in the *Skills Measured* area of *Describe an analytics workload on Azure* from the Microsoft exam study guide. This represents 25%–30% of the total *Skills Assessed*.

By the end of this chapter, you will have a comprehensive understanding of real-time data analytics solutions, equipping you with the knowledge to make informed decisions about data ingestion, processing, storage, and visualization options, and using Azure services in real-world scenarios. This skill set is increasingly important as organizations leverage real-time data to make decisions, thus attaining a competitive advantage in an ever-evolving marketplace.

Completing this chapter ensures you have the skills ready for the certification exam.

You will start this chapter by looking at the difference between batch and streaming data.

Describe the Difference between Batch and Streaming Data

Batch and streaming data are ways of processing data to fulfill organizations' different needs and use cases.

In short, the main difference between batch and streaming data is when the data is processed; batch processing happens periodically, at regular/ scheduled intervals, whereas streaming processes data instantaneously as it arrives in near-real time.

The following sections will explore these two ways to process data.

What Is Batch Processing?

Batch data is processed regularly in large batches, where data is collected and stored over a period of time for later processing as a batch (as a collective action).

Batch processing is used in processes where time-sensitive or immediate processing is unnecessary – for example, overnight processing at the end of the day or end-of-month processing for financial reports. Most of the time, data does not need to be structured or ordered in any sequence.

Batch processing can process large amounts of data quickly. It is a useful technique for performing tasks that require a massive amount of data, but the processing does not occur on the latest dataset; latency is expected when the data is processed.

Figure 7.1 illustrates the batch processing method. In this dining scenario, a tab is created and kept open for the dining period. As time progresses, additional items, such as food and drinks, are added as individual transaction records. Once the individual records of all these items have been stored, a single transaction can be executed to process all the items together at some later point when the bill is to be settled (i.e., all transactions recorded are batched and presented as one bill, not the individual records of items to pay).

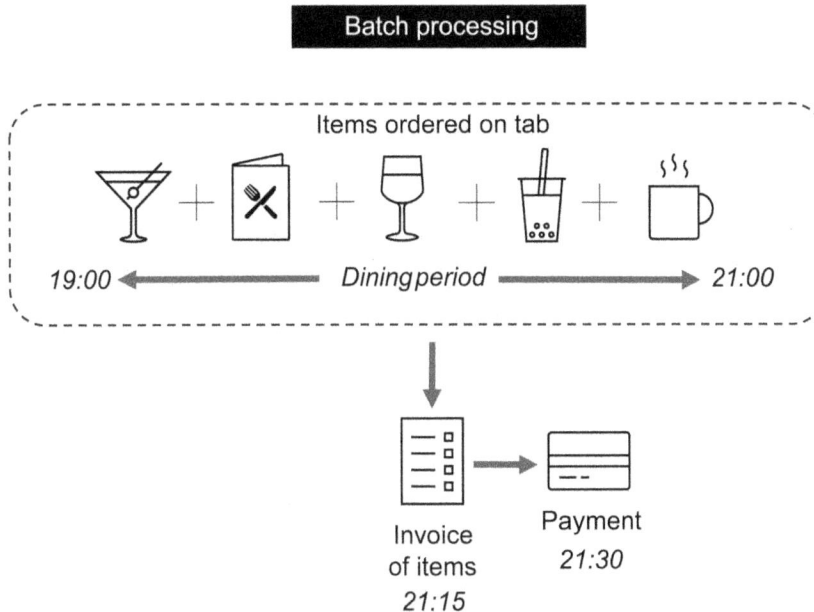

Figure 7.1: A batch processing example scenario

The advantage of the batch processing method can be seen from the scenario in *Figure 7.1*. All the individual order transactions of drinks and food can be collected and then processed efficiently, with one invoice and one payment transaction at a convenient time, during or at the end of the dining period (i.e., you could immediately ask for a bill when your coffee is ordered and make payment immediately, or you could wait some time till after you finish your coffee, then ask for the bill, and sometime later, when convenient, pay when you leave).

There is also a disadvantage to batch processing; before the batch can be run, all the items ordered on the tab for the dining period must be available to add to the final bill. However, this payment (batch task) could be stopped or not be possible if an item is missing or added incorrectly (i.e., if the bill is queried, it must be checked and amended before it can be settled, with the new "clean" correct data presented). Collecting payment is based on the newly batched dining period items presented to the customer for the payment task to complete.

Next, you will learn about stream processing and how this method contrasts and differs from batch processing.

What Is Stream Processing?

Stream processing refers to collecting (ingesting) and processing aggregations of data in real time as it is generated from various sources into a stream processing system, for real-time analysis.

Data stream processing is done continuously, with small and incremental updates being processed as soon as they are generated. The most recent data received is processed during the continuous time window; this data could be an IoT sensor that sends readings per second or social media feeds per minute.

The benefit of stream processing is that it allows organizations to act on data insights as soon as they are generated, with a rolling average calculated. A latency of seconds or milliseconds is required (expected) to process the streamed data immediately after the trigger.

Stream processing is used for real-time tasks, such as analyzing social media streams or monitoring network traffic for security threats.

Figure 7.2 illustrates the stream processing method. In contrast to the batch processing method illustrated in *Figure 7.1*, each item (record) ordered must be paid for immediately; there is no option to delay payment or batch the orders and make a single transaction payment for all items later, as you can do in batch processing. This is a pay-as-you-go approach where each item must be paid for (i.e., the record must be processed) in real time.

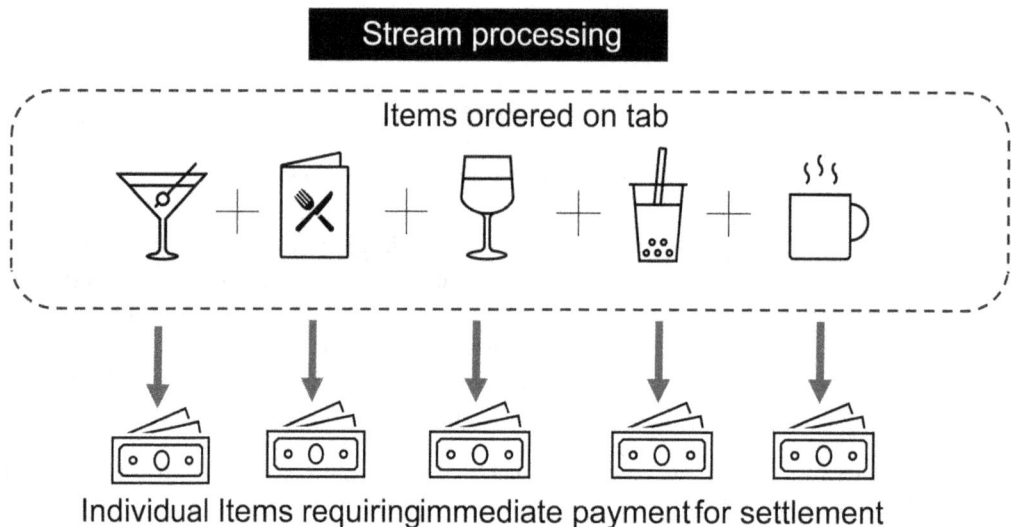

Figure 7.2: A stream processing example scenario

In this scenario, you get a real-time response when you wish to determine how much your payments are at any point in time, rather than waiting until the end of the dining period when all recorded items have been collected and calculated.

Some stream processing concepts that need to be understood are **sources** and **sinks**. We will explain what they are next.

Stream Processing Sources

Stream processing sources refer to where data is **ingested** (i.e., collected) from for further processing or storage. The common sources for stream processing ingestion will vary, depending on the use cases and industries; the following are some examples:

- **Messaging systems**: Large volumes of streaming data from sensors, social media, and web applications are ingested into messaging systems such as Apache Kafka and Azure Event Hubs

- **File systems**: Distributed file storage systems such as **Hadoop Distributed File System (HDFS)** can ingest and process large data volumes in real time

- **Data stores**: Operational databases such as SQL, Apache Cassandra, MongoDB, or NoSQL databases

- **APIs**: Applications with APIs that provide capabilities for ingestion from an analytics solution

- **IoT Hub**: IoT Hub is a service for ingesting and processing IoT device-generated data

- **Data integration platforms**: Azure Data Factory can pull in and transform streaming data from disparate sources, shaping it into a format that works for stream processing systems

These are only a few examples of various sources leveraging stream processing ingestion. Numerous use cases and industries exist for stream processing ingestion, such as real-time monitoring, fraud detection, anomaly detection, and recommendation engines.

Stream Processing Sinks

Sinks are the endpoints where the data stream is delivered for further processing or storage. Some of the commonly used sinks for stream processing outputs are as follows:

- **Databases**: The results of the stream processing can be fed into a relational or NoSQL database (e.g., Apache Cassandra, MongoDB, or Azure Cosmos DB) for further analysis or storage of the data, involving very large amounts of information, with excellent query performance

- **Analytical data (file) stores**: The analyzed data can be pushed to cloud-based data analytics stores (data warehouses and lakehouses) for analysis or reporting, using sophisticated **business intelligence (BI)** tools or query engines

- **Messaging systems**: Stream-processed data can be sent to messaging systems such as Apache Kafka or Azure Event Hubs for further processing or delivery to downstream systems

- **APIs**: The stream-processing system can output analyzed data to software applications for further processing or delivery to other downstream systems

- **Dashboards**: Processed Big Data can be routed to stream processing systems for real-time dashboards such as Power BI

These are just a few examples of the sinks for stream processing outputs. Of course, the sink to which you send data depends on the use case and industry. Stream processing systems can also serve multiple sinks parallel to support different downstream systems and use cases.

As you have seen, batch processing and stream processing are different ways of processing data and have specific advantages.

Organizations use batch and stream processing according to their needs and use cases.

Combining Stream and Batch Processing

Many organizations use a combination of batch and stream processing to manage their data effectively.

Some examples of organizations that could benefit from this combination are e-commerce platforms, retail chains, financial institutions, and healthcare providers.

You can benefit from both approaches by combining batch and stream processing. As you discovered in the previous sections, batch processing efficiently handles large volumes of data, while stream processing is ideal for real-time insights.

With the appropriate data warehouses, BI, or other data management tools, batch and stream processing can be integrated, processing both data methods on the same platform.

Figure 7.3 illustrates an example reference architecture that shows the data flow in a system that combines batch and stream processing for data analytics.

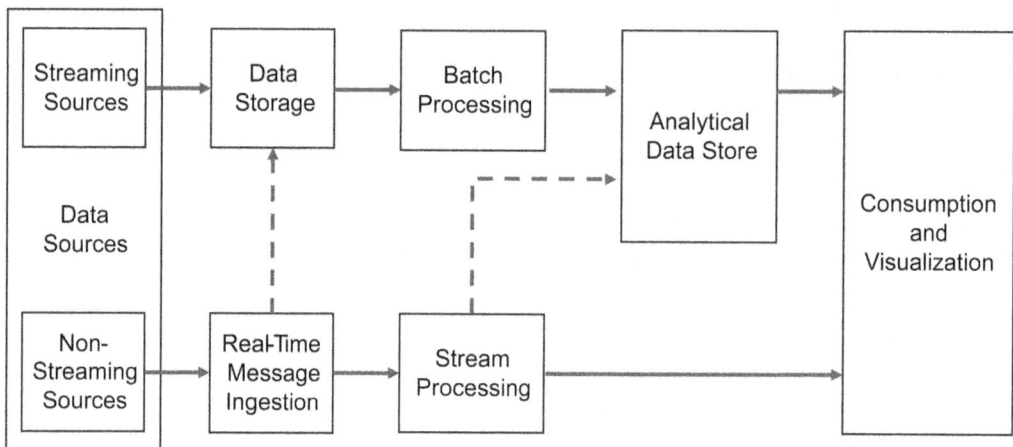

Figure 7.3: A combined stream and batch data processing architecture and flows

Here is an explanation of the diagram shown in *Figure 7.3*:

- **Data sources**: The system collects data from two types of sources – *streaming sources* (such as IoT devices, social media feeds, or online transactions) that generate continuous, real-time data, and *non-streaming sources* (such as databases, logs, or historical data) that produce data in batches or at specific intervals.

- **Data storage**: The data from streaming and non-streaming sources is first stored in a data storage system. This storage acts as a buffer or repository where data can be held before processing.

- **Batch processing**: Batch processing is applied to the data stored from non-streaming sources. This involves processing large volumes of data in scheduled intervals, often used for tasks such as data cleansing, aggregation, or historical analysis.

- **Real-time message ingestion**: Streaming data is immediately fed into a real-time message ingestion system, which prepares the data for stream processing. This could involve systems such as Apache Kafka or Azure Event Hubs, which handle real-time data's high throughput and low-latency requirements.

- **Stream processing**: In this stage, stream processing handles the real-time data as it arrives, allowing for immediate analysis, filtering, and pattern recognition. This is crucial for fraud detection, real-time recommendations, and live monitoring applications.

- **Analytical data store**: Both batch-processed and stream-processed data converge into an analytical data store. This centralized repository holds the processed data, making it ready for further analysis or visualization. Depending on an organization's needs, this store might be a data warehouse or a data lake.

- **Consumption and visualization**: The final stage is consumption and visualization, where the processed data is used for reporting, dashboarding, or feeding into BI tools. This stage allows stakeholders to gain insights, make data-driven decisions, and visualize trends or anomalies in the data.

Large amounts of data can be processed using batch and stream processing. By doing this, you can help a company reach its decision by looking deeply into its data.

> **Note**
>
> Batch and stream processing solutions commonly use a **lambda** or **kappa** architecture, which are beyond the scope of this content; more information can be found at `https://learn.microsoft.com/en-us/azure/architecture/databases/guide/big-data-architectures`.

Identify Microsoft Cloud Services for Real-Time Analytics

Microsoft offers a range of real-time analytics services to process and analyze large volumes of streaming data as close as possible to real time, leading to actionable insights, anomaly detection, alerts, and responses.

Microsoft's real-time analytics services include the following:

- **Azure Stream Analytics**: With this service, you analyze real-time data using simple queries, written in a SQL-like language. You can use this service with many input and output sources, and real-time monitoring and autoscaling are built in.

- **Spark Structured Streaming**: This stream processing engine on top of Apache Spark SQL allows you to use a high-level API to process data streams in real time, in a distributed, scalable, and fault-tolerant manner.

- **Azure Data Explorer**: A massively scalable data analytics service designed for the fast querying and visualization of large volumes of telemetry data, logs, and time-series data (queries, or KQL, are executed against tables for querying). It can be run as a standalone service or in a Synapse workspace for Azure Synapse Analytics, as an Azure Synapse Data Explorer runtime.

- **Microsoft Fabric**: This **software-as-a-service** (**SaaS**) analytics platform provides real-time data analytics using multiple sources of continuous data ingestion. Microsoft Fabric uses Azure Synapse Analytics as its compute service, while Eventstream is used to persist ingested real-time streaming source data in a lake house or KQL database table. The real-time data can be queried using SQL and KQL.

These solutions can be applied in the real-time processing of IoT data, real-time monitoring of social media feeds, real-time analysis of customer feedback, and real-time detection of anomalies (including automated data breaches) in application logs.

The following are some common **sources** (inputs) available on Azure for stream processing:

- **Apache Kafka**
- **Azure Data Lake Store Gen 2**
- **Azure Event Hubs**
- **Azure IoT Hub**

The following are common **sinks** (destinations) available on Azure for stream processing:

- **Azure Data Lake Store Gen 2**
- **Azure Blob Storage**
- **Azure SQL Database**
- **Azure Synapse Analytics**
- **Azure Databricks**
- **Azure Event Hubs**
- **Power BI**

You will learn about Microsoft real-time analytics solutions in the following sections.

Azure Stream Analytics

Azure Stream Analytics works to enable the processing and analyzing of data originating from streaming sources at a high volume and in real time, enabling enterprises to respond to events proactively, as determined by the incoming insights, and try to derive more insights from the streaming data, which is continuously changing or evolving. Azure Stream Analytics can take data from sources as diverse as the **Internet of Things (IoT)**, social media, and log files.

Dashboards can be created from the insights extracted from the data streams to make real-time decisions.

Azure Stream Analytics is also tightly integrated with other Azure services, such as Azure Event Hubs and Azure IoT Hub, to make it easy for a business to pull data from multiple input sources.

Apache Spark

Apache Spark is a single unified analytics engine for big data processing workloads and heterogeneous environments, including structured data, unstructured data, graph data, streaming data, and multiple languages for application development, all wrapped in a single analytics engine for high performance, resource utilization efficiency, and ease of use. Apache Spark supports workloads with high throughput, low latency, and high scalability, as well as applications written in Java, Scala, Python, or R.

At its core, Apache Spark accesses data stored in the HDFS. It can compute data held in memory and potentially be around 100 times faster than processing the same volume of data with Hadoop's MapReduce framework. Moreover, it includes libraries for machine learning, graph processing, and processing streamed data, making it a new big data processing platform capable of processing many different kinds of data.

On top of this, Apache Spark exposes a higher-level abstraction on top of the dataset, known to developers as the **DataFrame**. The DataFrame API is a higher-level abstraction that makes it easier for developers to work with data in a structured format – it allows developers to manipulate data as if it were via SQL-like queries (including joins, `group by` statements, aggregations, and string manipulation operations). Developers can also treat data as tabular data and manipulate it as such (such as adding or removing columns to the data). Because of the underlying parallelism in Apache Spark, the DataFrame API can efficiently scale across multiple machines, enabling developers to do real-time analyses on large and massive data.

Apache Spark is extremely fast and robust against faults. It has an incredible ecosystem of additional libraries, making it easy to process lots of data in parallel, including streaming data (for example, from a website's traffic) and sophisticated machine learning. Apache Spark fault tolerance means that processing jobs are unaffected by falling computers on a cluster.

Azure Data Explorer

This data analytics service is provided by Microsoft Azure; it enables real-time processing, analysis, and visualization of big data.

The service uses the Microsoft **Kusto Query Language (KQL)** for big data querying and analysis. It enables end users to issue a proprietary SQL-like query language, capable of running complex operations, aggregations, and joining to ease querying and analyzing large datasets. The query language is gamed for scalable analysis of ingested data streams, enabling response in near-real time.

Real-time data ingestion and analysis are significant features of Azure Data Explorer, with dashboards that can be created from insight derived from the data streams, allowing a user to respond dynamically.

Microsoft Fabric

Microsoft Fabric is a SaaS, cloud-based, data analytics solution that helps diverse data personas across a business to ingest, transform, analyze, and visualize data, end to end, from different data sources in the same browser-based experience.

Users can uncover meaningful insights at their fingertips by exploiting the analytics, data integration, and visualization capabilities of this versatile, all-round, purpose-built unified analytics platform. Its highly user-friendly and usable persona-driven design, as well as myriads of tools and features, make it a compelling choice to enable well-informed, data-inspired decision-making.

This section on Microsoft Fabric concludes our discussion of real-time analytics with Microsoft services. Now, it is time to summarize what skills you have learned in this chapter.

Summary

This chapter included complete coverage of the *DP-900 Azure Data Fundamentals* exam's *Describe Consideration for Real-Time Data Analytics* skills measured area.

In this chapter, you gained a comprehensive understanding of real-time data analytics, focusing on the fundamental differences between batch and streaming data processing and the key Azure services that support real-time analytics.

You began by identifying the difference between batch and streaming data processing. Batch processing involves collecting and processing large volumes of data at scheduled intervals, making it ideal for tasks where immediate processing is unnecessary, such as end-of-day financial reporting. In contrast, streaming data processing handles data in real time, processing it as it arrives. It is crucial for applications that require immediate insights, such as monitoring network traffic or analyzing social media feeds.

The chapter also introduced you to the various Microsoft Azure data platform services that enable real-time analytics, including Azure Stream Analytics, Apache Spark on Azure, Azure Data Explorer, and Microsoft Fabric.

Additionally, we discussed how these Azure data platform services integrate with other tools and services, offering flexibility and scalability to meet different data processing needs. A knowledge of these tools allows you to choose the right service for specific real-time analytics tasks, whether for IoT data ingestion, anomaly detection, or real-time dashboarding.

By understanding the differences between batch and stream processing and familiarizing yourself with Azure's real-time analytics services, you are now better equipped to implement effective data analytics solutions. These skills are crucial for passing the DP-900 certification exam and addressing real-world business challenges, where timely and accurate data insights can drive strategic decision-making.

The next chapter, *Describe Data Visualization in Microsoft Power BI*, will follow the skills measured sequence in Microsoft's study guide for the DP-900 exam.

Additional Reading

This section provides links to additional exam information and study references:

- DP-900 - Microsoft Azure Data Fundamentals study guide: `https://learn.microsoft.com/en-us/credentials/certifications/resources/study-guides/dp-900`

- DP-900 - Microsoft Azure Data Fundamentals self-directed learning: `https://learn.microsoft.com/en-gb/training/modules/explore-fundamentals-stream-processing/`

Exam Readiness Drill – Chapter Review Questions

Apart from a solid understanding of key concepts, being able to think quickly under time pressure is a skill that will help you ace your certification exam. That is why working on these skills early on in your learning journey is key.

Chapter review questions are designed to improve your test-taking skills progressively with each chapter you learn and review your understanding of key concepts in the chapter at the same time. You'll find these at the end of each chapter.

> **How to Access These Materials**
>
> To learn how to access these resources, head over to the chapter titled *Chapter 9, Accessing the Online Resources*.

To open the Chapter Review Questions for this chapter, perform the following steps:

1. Click the link – `https://packt.link/DP900Ch07`.

   Alternatively, you can scan the following **QR code** (*Figure 7.4*):

Figure 7.4: QR code that opens Chapter Review Questions for logged-in users

2. Once you log in, you'll see a page similar to the one shown in *Figure 7.5*:

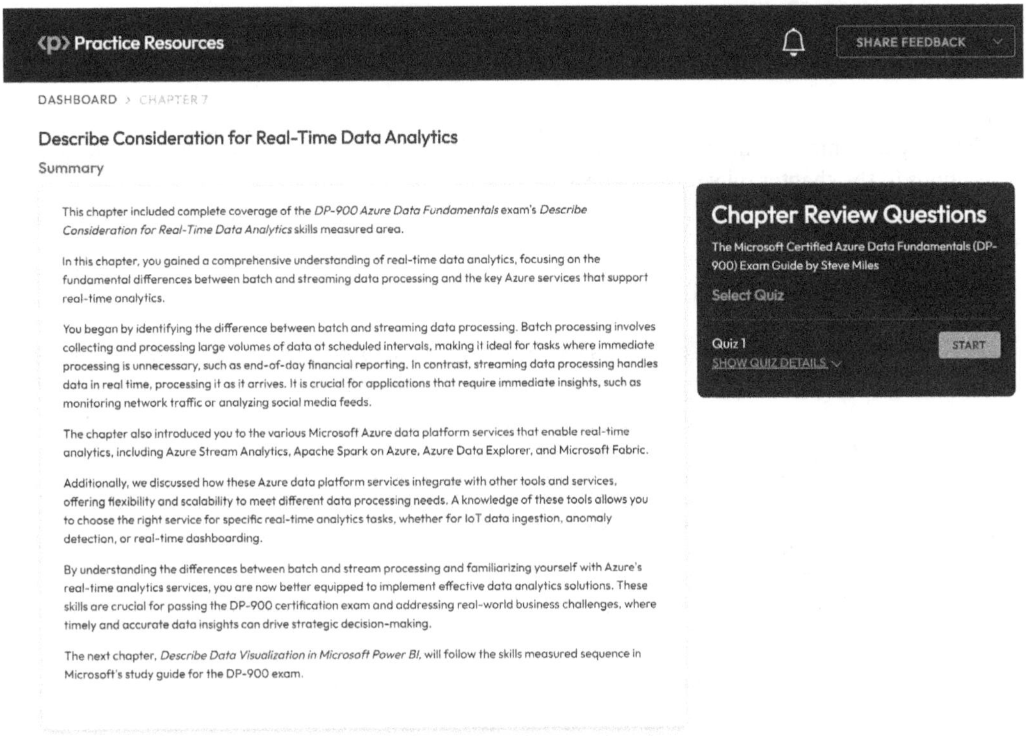

Figure 7.5: Chapter Review Questions for Chapter 7

3. Once ready, start the following practice drills, re-attempting the quiz multiple times.

Exam Readiness Drill

For the first three attempts, don't worry about the time limit.

ATTEMPT 1

The first time, aim for at least **40%**. Look at the answers you got wrong and read the relevant sections in the chapter again to fix your learning gaps.

ATTEMPT 2

The second time, aim for at least **60%**. Look at the answers you got wrong and read the relevant sections in the chapter again to fix any remaining learning gaps.

ATTEMPT 3

The third time, aim for at least **75%**. Once you score 75% or more, you start working on your timing.

> **Tip**
> You may take more than **three** attempts to reach 75%. That's okay. Just review the relevant sections in the chapter till you get there.

Working On Timing

Target: Your aim is to keep the score the same while trying to answer these questions as quickly as possible. Here's an example of how your next attempts should look like:

Attempt	Score	Time Taken
Attempt 5	77%	21 mins 30 seconds
Attempt 6	78%	18 mins 34 seconds
Attempt 7	76%	14 mins 44 seconds

Table 7.1: Sample timing practice drills on the online platform

> **Note**
> The time limits shown in the above table are just examples. Set your own time limits with each attempt based on the time limit of the quiz on the website.

With each new attempt, your score should stay above **75%** while your "time taken" to complete should "decrease". Repeat as many attempts as you want till you feel confident dealing with the time pressure.

Describe Data Visualization in Microsoft BI

This chapter will explore the powerful data visualization capabilities in **Microsoft Power BI**, a leading business intelligence tool in the Microsoft ecosystem.

Data visualization is a critical component of data analytics, transforming raw data into visual formats such as charts, graphs, and dashboards that make complex information more accessible and actionable for decision-makers.

This chapter includes learning content for the *DP-900 Azure Data Fundamentals exam*, focusing on the *Skills Measured* area of *Describe an analytics workload on Azure* from the *Microsoft exam study guide*. This represents 25%–30% of the total *Skills Assessed*.

You will begin by discussing the importance of data visualization and how it enhances the comprehension and communication of data insights. Effective data visualization not only helps in identifying patterns and trends but also enables users to make informed decisions quickly, by presenting data in a visually compelling way.

This chapter will also cover the key features and tools within Microsoft Power BI that facilitate data visualization. You will examine interactive reports, dashboards, and custom visualizations to meet specific business needs and share insights across an organization. Additionally, you will explore how Power BI integrates with other Microsoft Azure data platform services, enabling seamless data ingestion, processing, and visualization in a unified environment.

In this chapter, we will cover the following main topics:

- Identify the capabilities of Power BI
- Describe the core concepts of data modeling
- Describe the features of data models in Power BI
- Identify the appropriate visualizations for data

Understanding data visualization in Microsoft Power BI is essential for leveraging the full potential of your data, whether you are analyzing sales trends, monitoring operational performance, or exploring customer behavior. By the end of this chapter, you will have the knowledge and skills to create impactful visualizations that drive data-driven decision-making, positioning you to effectively use Power BI in real-world scenarios and prepare for related certification exams.

The first section of this chapter will teach you how to identify Power BI capabilities.

Identify the Capabilities of Power BI

Data visualization is an integral component of any **data analytics solution**.

Data visualization makes it easier to look at vast amounts of data and identify patterns, trends, and relationships that might not be obvious. Interactive and engaging presentation of the data focuses the attention of stakeholders on its significance, meaning, and the importance of the insights that can be gathered from it. This process leads to better-informed decision-making and increased understanding, allowing organizations to make decisions about factors that will generate growth, improved performance, and customer satisfaction.

Consider the sales data of thousands of products across multiple regions and stores over several months. When this data is stored in spreadsheets or databases, it's not easy to spot patterns, trends, or outliers. Important details such as which products are performing well, which regions are lagging, or which time periods see the most sales might be hidden in the sheer volume of numbers.

This is where data visualization in Power BI comes into play. By transforming raw data into visual formats (such as interactive charts, graphs, and dashboards), Power BI makes it possible to quickly and easily identify trends, correlations, and anomalies. For example, a sales manager can instantly see which products are top sellers in specific regions through a bar chart or identify seasonal trends in sales with a line graph. Instead of sifting through rows of numbers, stakeholders can visually comprehend the story that the data tells, leading to quicker, more informed decision-making.

Power BI's capabilities go beyond basic visualizations; it allows users to create custom visuals, drill down into data for deeper analysis, and predict future trends with advanced analytics. This not only helps in making sense of vast amounts of data but also in communicating insights effectively across an organization.

Microsoft Power BI is a visualization tool that allows users to create an all-in-one business intelligence and reporting solution.

Figure 8.1 outlines the architecture and process of creating and consuming data visualizations using Power BI:

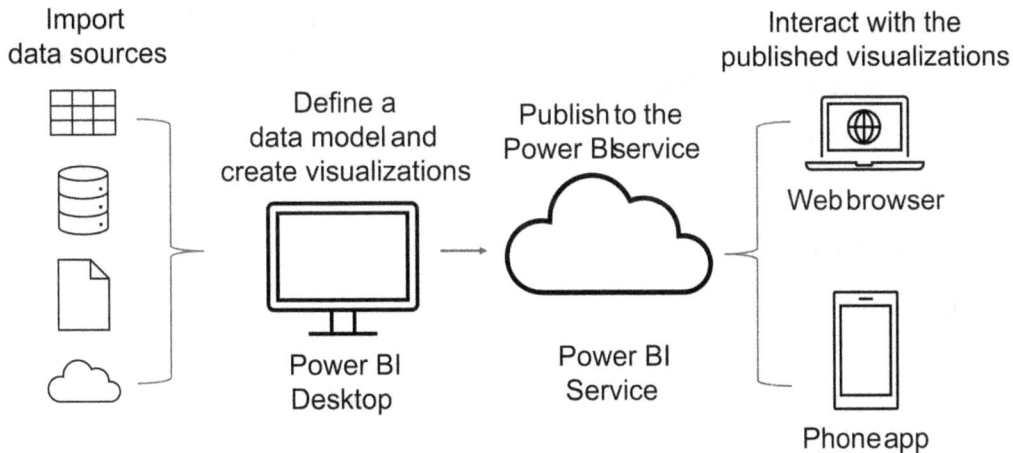

Figure 8.1: The Power BI architecture

As you can see from *Figure 8.1*, the Power BI suite includes two core components – **Power BI Desktop** and the **Power BI Service**.

Some of the key capabilities of Power BI are as follows:

* **Data visualization**: Power BI provides data visualizations in interactive and visually engaging formats, including charts and graphs, with presentations that can be viewed as collections in dashboards and reports created by analysts/report writers. You can create reports, which are multi-page documents that provide detailed analysis through various visualizations, and dashboards, which are single-page, high-level overviews that consolidate key metrics from different reports for quick insights.

* **Data modeling**: It provides the ability to create data models to help organize and analyze data.

* **Data analysis**: When combined with pivot tables, DAX formulas, and quick measures, Power BI enables users to perform complex data analysis.

* **Data connectivity**: Various data sources such as Excel spreadsheets, SQL Server databases, or cloud-based data such as Azure and AWS can be connected.

* **Collaboration**: Real-time reports and dashboards are shared with multiple users and business units, which enables two or more teams to work together.

* **Mobile support**: Power BI allows users to access reports and dashboards anywhere and on any device. Power BI also offers Power BI apps, which enhance mobile support by allowing users to easily access, share, and interact with their reports and dashboards on any device, ensuring that critical insights are always at their fingertips.

The benefits of data visualization are that it allows users to perform analyses, make predictions, and estimate impacts in real time. Simultaneously, it allows organizations to make data-informed decisions based on the most pertinent information.

The core Power BI Desktop and Power BI service components shown in *Figure 8.1* are covered in the following sections.

Power BI Desktop

With **Power BI Desktop**, you can import and prepare data from different sources, design and edit reports and visualizations, and create and edit reports and visualizations.

With Power BI Desktop as the client application, data modeling (structuring and organizing data) and analysis are executed on the local machine. Data transformation (cleaning, shaping, and modifying raw data to make it suitable for analysis) can be done using the dataflows, relationships can be built between tables, and calculated columns and measures can be created. Different visualizations support tables, charts, maps, and gauges, among many others. Reports and dashboards can be designed in Power BI Desktop, generating many custom and interactive reports and dashboards. Drilling down and some "what-if" scenario analyses can also be performed.

In Power BI, "drilling down" refers to exploring data at different levels of detail within a report or dashboard. When you drill down, you move from a summary view to more detailed data, often within a hierarchy. For example, in this type of analysis, if you're looking at sales data, you might start with a high-level view showing total sales by year. Drilling down could then show sales by quarter, then by month, and finally, by individual day. This feature allows users to analyze data at various granularities, helping them uncover insights that aren't immediately visible at the summary level.

Power BI Desktop visualizations comprise the following methods to present information:

- Reports
- Dashboards

You will learn more about these in the following two sections.

Dashboards

Dashboards are a Power BI feature that allows users to provide a one-page view that contains visualizations and communicates important data points. These views provide a focused perspective of critical information from different data sources and allow end users to monitor the most important metrics at a glance. Each workspace has a dedicated dashboard associated with it.

A dashboard can include a variety of data visualizations, such as charts, graphs, and tables, to give an interactive and understandable overview of all data. The dashboard can utilize one or more visualizations from different datasets/reports, databases, and spreadsheets.

Reports

Reports typically comprise several pages of visualization: the body of the report itself may include interactive **drill-down** visuals, such as bar/column charts, pie charts, maps, tables/matrices, and the output of prebuilt queries.

For instance, the drill-down functionality explained earlier in this chapter is an amazing feature in Power BI that allows users to work interactively on a dataset at different levels of detail. If hierarchical data is set up and drill-down functionality is enabled in the visualizations, users can perform analysis at different levels of granularity without having to create dozens of independent reports.

Power BI Report Builder

Report Builder is not a Power BI Desktop component but a standalone service that is installed and used to create paginated reports.

These types of reports are good for cases where they need to be **highly formatted** and have **fixed layouts**, such as financial statements, invoices, forms, or operational reports. **Paginated** reports can span multiple pages, with **precise pixel control** over headers, footers, and page breaks, among other things. You can save the reports and upload them to a Power BI Service workspace that is shared with others.

The Power BI service is covered in the following section.

The Power BI Service

The **Power BI service** is a cloud-based platform that enables the publishing, sharing, and collaboration of Power BI reports and dashboards, as well as the management and utilization of semantic models.

It enables a centrally managed, shared environment for publishing and collaborating on reports that can be used by a whole team working on a data analysis project. It also offers natural language queries and Q&A capabilities, as well as mobile access to data so that you can access and interact with it wherever you are.

Better Together

When Power BI Desktop is used, Power Query Editor allows users to perform complex queries and manipulate data. Simultaneously, the Power View element enables the creation of a panoramic view of the reports. Power BI Desktop and the Power BI service create an integrated end-to-end data analysis and reporting environment.

The difference is that these components offer different capabilities; there's a separation of concerns between the authoring capabilities provided by Power BI Desktop and the consumption/sharing/collaboration capabilities provided by the Power BI service, which provides the cloud service sharing capability, as well as a refresh and auto-update of data both on demand and on schedule.

You have now learned that Power BI Desktop and the Power BI service offer users an option of business intelligence tools to help analyze data and report business insights. By being able to build these reports and dashboards in Power BI Desktop, users can easily publish these reports to share with colleagues or even external parties whom our company wants to report.

Now that you have learned the capabilities of Power BI, you will be taught the features of Power BI data models.

Describe Core Concepts of Data Modeling

Analytical models give you a structure for the data you want to analyze. In the following sections, we will explore some core concepts, such as **cubes** and **schemas**.

Cubes

Analytical models are based on related data tables that contain the following properties:

- **Measures**: These are the **numeric values** for a model; they can include sales metrics such as quantity or revenue
- **Dimensions**: These are descriptive attributes for a specific aspect of data, such as customer, product, and time

A typical data table in a business context often contains quantitative sales metrics (measures), such as revenue or quantity, and categorical dimensions, such as product, customer, and time. With this data, analysts can perform in-depth analyses to derive insights, such as the total revenue by an individual customer or the monthly number of items sold by a product.

Conceptually, a model becomes a **multidimensional cube**, and at any point at which the dimensions intersect, we have an **aggregate value** (*aggregate measure*) for those particular dimensions.

Figure 8.2 represents the **analytical model** (cube) concept:

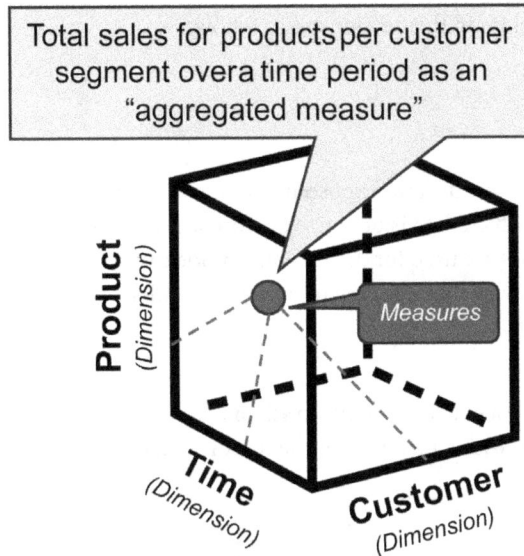

Figure 8.2: An analytical model showing dimensions and measures

The illustration in *Figure 8.2* is a 3D cube representing a multidimensional data model. The three axes of the cube are labeled as **Product**, **Time**, and **Customer**. These are the dimensions of the data. The center of the cube contains a blue dot labeled **Measures**. This represents the **aggregated metric** of the total sales of products sold to customers over time. The dashed lines extending from the blue dot to the three axes indicate that the aggregated sales data is analyzed across all three dimensions.

This visualization is a helpful tool for understanding how different factors can impact sales performance. For example, by analyzing sales data across different products, time periods, and customer segments, businesses can identify which products are selling well, when sales are typically highest, and who their best customers are. This information can then be used to make informed decisions about marketing, product development, and sales strategies.

> **Note**
> There can be more than three dimensions in an analytical cube, but this is hard to represent and visualize, so explaining concepts is the simplest way to understand the model.

Schemas

Schemas are fundamental database designs used in data warehousing and business intelligence applications. These schemas are essential for efficiently organizing and analyzing large datasets, providing a structured foundation for data-driven decision-making.

The schemas you will explore in this section are as follows:

- Star schema

- Snowflake schema

Star and **snowflake** schemas are data warehouse and business intelligence database schemas. A **database schema** is a structure by which data is organized before it actually resides within the database. Both star and snowflake schemas are used for dimensional modeling, a technique in data warehousing for structuring data in such a way that it is easy to query and understand.

Star Schema

A **star schema** is the simplest entity-relationship schema for a data warehouse. It's called a star schema because it looks like a star, with a fact table in the middle and dimension tables radiating from it.

Figure 8.3 illustrates the star schema:

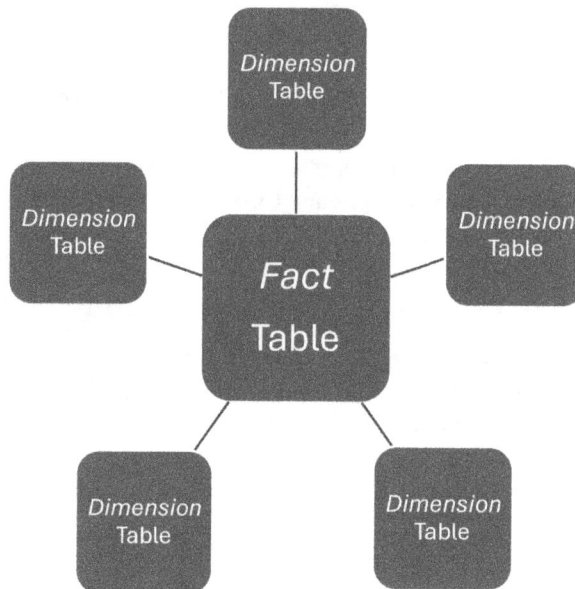

Figure 8.3: A star schema

Now, take a look at the key components of a star schema. They are as follows:

- **Fact table**: This central table contains a business's measurable, quantitative data (facts). It includes keys to the dimension tables and numeric metrics such as sales amount and order quantity.

- **Dimension tables**: These tables contain descriptive attributes (dimensions) related to the facts. Dimension tables are typically denormalized and include columns such as `Product name`, `Customer name`, and `Period`.

The Advantages of a Star Schema

Star schemas offer several advantages for data warehousing and business intelligence applications. Their straightforward join paths between fact and dimension tables simplify queries, leading to faster execution. The denormalized structure with a central fact table and directly connected dimension tables improves query performance. Additionally, the schema's simplicity makes it easier for business users to understand and interpret data, facilitating data-driven decision-making.

Snowflake Schema

The **snowflake schema** enhances the star schema by normalizing the dimension tables to create multiple related tables.

Figure 8.4 illustrates the snowflake schema:

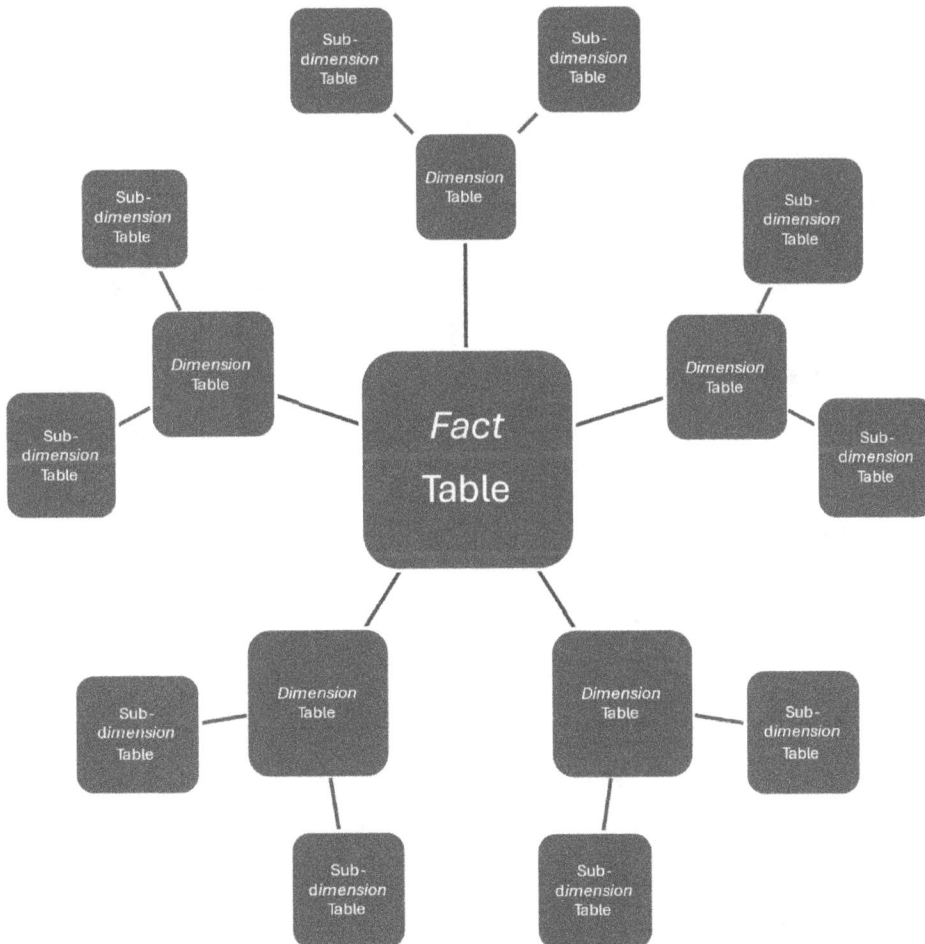

Figure 8.4: A snowflake schema

The name snowflake schema is derived from its visual resemblance to a snowflake.

The core structure is like a star schema; the snowflake schema begins with a central fact table.

However, unlike the star schema, dimension tables in the snowflake schema are further normalized. This means they are broken down into smaller sub-dimension tables, connected through foreign key relationships.

When visualized, these interconnected dimension tables create a branching pattern that resembles a snowflake's intricate design.

In essence, the snowflake schema's name reflects its hierarchical structure and the branching relationships between its dimension tables.

The key components of a snowflake schema are:

- **Fact table**: Similar to the star schema, it contains quantitative data and keys to dimension tables.

- **Normalized dimension tables**: Dimension tables in a snowflake schema are normalized. This means they are split into additional tables, reducing data redundancy. Each dimension can have multiple related tables, forming a structure resembling a snowflake.

The Advantages of a Snowflake Schema

The snowflake schema's normalized dimension tables effectively reduce data redundancy, leading to more efficient storage. Organizations can save storage space and improve data management by minimizing duplicate data.

The Disadvantages of a Snowflake Schema

While the snowflake schema offers benefits in terms of data redundancy, it can introduce complexities into query execution. The increased number of joins required to access data across multiple dimension tables can slow down query performance. This is especially noticeable when dealing with large datasets or complex analytical queries.

As you have seen, a snowflake schema is a more complex, normalized structure, with dimension tables split into multiple related tables. It reduces redundancy but may result in more complex and slower queries.

The final aspect of data models to explore is **hierarchies**; these represent the relationships between different levels of attributes within a **dimension**. For example, in a time dimension, a hierarchy might exist between the year, quarter, month, and date, where each level represents a more granular subdivision of time.

Within a hierarchy, there are individual layers or stages, which are referred to as **levels**. For instance, in a geographic dimension, levels might include the country, region, state, and city. Each level represents a different degree of granularity or detail that can be used for a drill-up or drill-down of a dimension.

This section explored the core data modeling concepts such as cubes and schemas. The next section will cover Power BI data models.

Describe the Features of Data Models in Power BI

Power BI has built-in data modeling tools that allow you to create complex data models with hierarchical, one-to-many, and many-to-many relationships.

By utilizing the `Data modeling` pane on the `Model` tab of Power BI Desktop, you can create and manage relationships between your fact and dimension tables, define hierarchies, define data types and display formats for tables and fields, and manage other properties for your data that define important aspects of your model for analysis.

In closing this section, you can now create a data model using Power BI, based on data tables extracted from one or more data sources.

This section explored Power BI data models. It is time to head to the end of this chapter by looking at the last topic, data visualizations, in the final section.

Identify Appropriate Visualizations for Data

Once a model has been implemented, reports can be created that include the data visualizations generated.

Power BI allows users to create these customized visualizations using built-in capabilities, such as **charts**, **graphs**, **maps**, and **tables**. These allow users to understand the significance of data and communicate insights effectively; they can provide patterns and trends and allow you to make data-driven decisions.

The following are some common visualizations that are used in reports and dashboards:

- **Table**: These are the easiest methods to represent data. Tables are very useful when lots of values are related to each other.
- **Bar and column chart**: These include at least two fixed values associated with discrete categories. They are generally used to support visual comparisons of discrete numeric values.
- **Line chart**: These can be used to compare categorized values and are useful when depicting how things change or examining trends, often over time.
- **Pie chart**: These are useful for comparing numerical values that are grouped in categories as a percentage of the total in business reports.

- **Scatter plot**: These are best when you want to compare two different measures of numbers and see whether there's a relationship between them or a correlation.

- **Map**: Maps enable you to compare the values measured in a geographic location or area.

Figure 8.5 represents these common data visualizations:

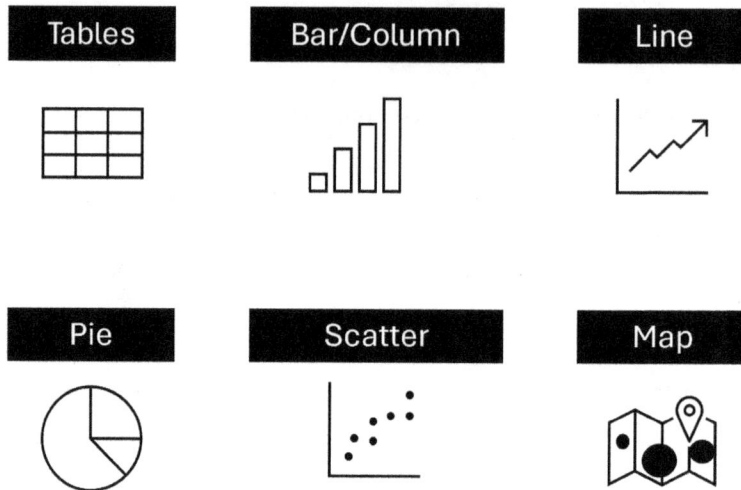

Figure 8.5: Common data visualizations

In conclusion, with Power BI, you can uncover hidden insights, identify new opportunities, and optimize your operations for greater efficiency and profitability.

This section on visualization for data concludes this chapter's content and the book's learning content. Now, it is time to summarize the skills you have learned in this chapter.

Summary

In this chapter, you learned about the capabilities of Power BI, the features of data models in Power BI, and appropriate visualizations for data. This chapter included complete coverage of the *DP-900 Azure Data Fundamentals* exam's *Skills Measured* area – *Describe data visualization in Microsoft Power BI*.

The chapter highlighted the importance of data visualization in enhancing the comprehension and communication of complex information. By transforming raw data into visual formats such as charts, graphs, and dashboards, Power BI helps decision-makers quickly identify patterns, trends, and actionable insights, making data more accessible and meaningful.

The chapter examined Power BI's features in detail, focusing on its core components – Power BI Desktop and the Power BI service. These tools work together to create a comprehensive business intelligence solution. The capabilities of Power BI were discussed, including its ability to connect with various data sources, model data effectively, and facilitate real-time collaboration through interactive reports and dashboards. This discussion emphasized how Power BI enables users to analyze data and share insights across an organization, making it a versatile tool for any data-driven enterprise.

You were also taught an understanding of fundamental data modeling concepts, such as analytical models, star schemas, and snowflake schemas. These concepts are essential for structuring data to optimize query performance and enhance the ease of understanding. The chapter illustrated how these models form the backbone of data organization within Power BI, enabling efficient and effective data analysis.

The chapter also explored the appropriate use of different data visualizations, guiding you through selecting visual formats that best represent various kinds of data. This knowledge is crucial for creating visualizations that look appealing and effectively communicate the underlying data's significance.

One of the major takeaways from this chapter is the critical role data visualization plays in decision-making. Effective visualization is more than aesthetics; it's also about making data understandable and actionable. Well-designed visualizations can significantly influence business decisions by presenting critical information clearly and concisely.

Additional Reading

This section provides links to additional exam information and study references:

- DP-900 - Microsoft Azure Data Fundamentals study guide: `https://learn.microsoft.com/en-us/credentials/certifications/resources/study-guides/dp-900`

- DP-900 - Microsoft Azure Data Fundamentals self-directed learning: `https://learn.microsoft.com/en-gb/training/modules/explore-fundamentals-data-visualization/`

Exam Readiness Drill – Chapter Review Questions

Apart from a solid understanding of key concepts, being able to think quickly under time pressure is a skill that will help you ace your certification exam. That is why working on these skills early on in your learning journey is key.

Chapter review questions are designed to improve your test-taking skills progressively with each chapter you learn and review your understanding of key concepts in the chapter at the same time. You'll find these at the end of each chapter.

> **How to Access These Materials**
>
> To learn how to access these resources, head over to the chapter titled *Chapter 9, Accessing the Online Resources.*

To open the Chapter Review Questions for this chapter, perform the following steps:

1. Click the link – `https://packt.link/DP900Ch08`.

   Alternatively, you can scan the following **QR code** (*Figure 8.6*):

Figure 8.6: QR code that opens Chapter Review Questions for logged-in users

2. Once you log in, you'll see a page similar to the one shown in *Figure 8.7*:

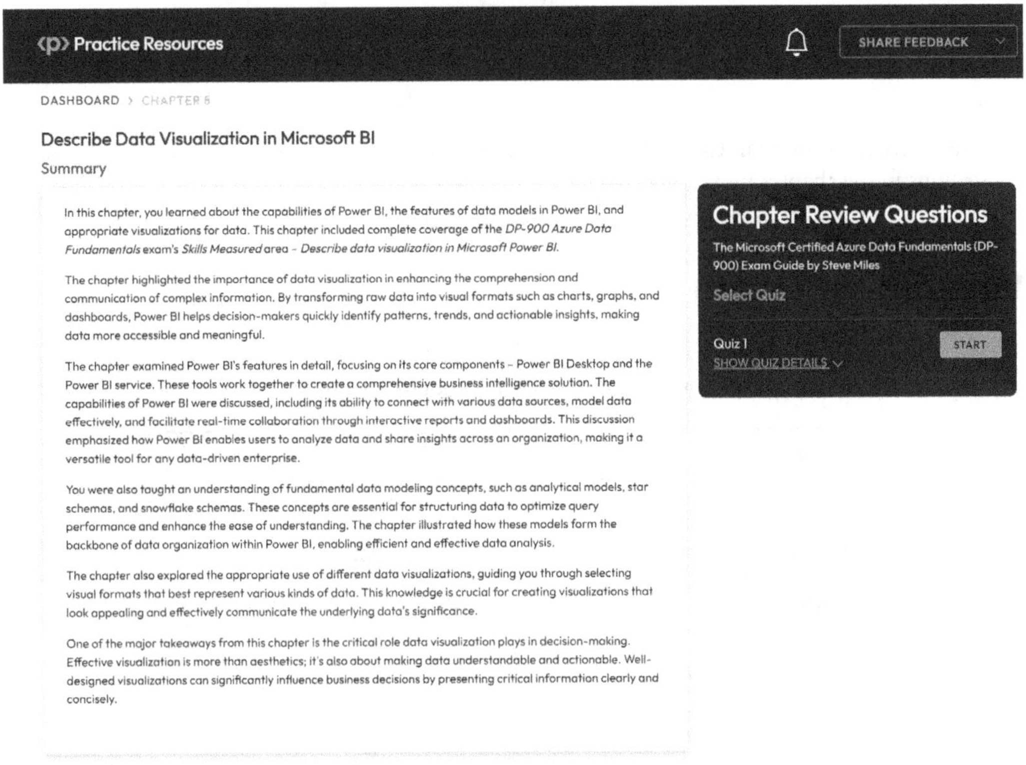

Figure 8.7: Chapter Review Questions for Chapter 8

3. Once ready, start the following practice drills, re-attempting the quiz multiple times.

Exam Readiness Drill

For the first three attempts, don't worry about the time limit.

ATTEMPT 1

The first time, aim for at least **40%**. Look at the answers you got wrong and read the relevant sections in the chapter again to fix your learning gaps.

ATTEMPT 2

The second time, aim for at least **60%**. Look at the answers you got wrong and read the relevant sections in the chapter again to fix any remaining learning gaps.

ATTEMPT 3

The third time, aim for at least **75%**. Once you score 75% or more, you start working on your timing.

> **Tip**
> You may take more than **three** attempts to reach 75%. That's okay. Just review the relevant sections in the chapter till you get there.

Working On Timing

Target: Your aim is to keep the score the same while trying to answer these questions as quickly as possible. Here's an example of how your next attempts should look like:

Attempt	Score	Time Taken
Attempt 5	77%	21 mins 30 seconds
Attempt 6	78%	18 mins 34 seconds
Attempt 7	76%	14 mins 44 seconds

Table 8.1: Sample timing practice drills on the online platform

> **Note**
> The time limits shown in the above table are just examples. Set your own time limits with each attempt based on the time limit of the quiz on the website.

With each new attempt, your score should stay above **75%** while your "time taken" to complete should "decrease". Repeat as many attempts as you want till you feel confident dealing with the time pressure.

Accessing the Online Practice Resources

9

Your copy of *Microsoft Certified Azure Data Fundamentals (DP-900) Exam Guide* comes with free online practice resources. Use these to hone your exam readiness even further by attempting practice questions on the companion website. The website is user-friendly and can be accessed from mobile, desktop, and tablet devices. It also includes interactive timers for an exam-like experience.

How to Access These Materials

Here's how you can start accessing these resources depending on your source of purchase.

Purchased from Packt Store (packtpub.com)

If you've bought the book from the Packt store (`packtpub.com`) eBook or Print, head to `https://packt.link/DP900UnlockResources`. There, log in using the same Packt account you created or used to purchase the book.

Packt+ Subscription

If you're a *Packt+ subscriber*, you can head over to the same link (`https://packt.link/DP900UnlockResources`), log in with your `Packt ID`, and start using the resources. You will have access to them as long as your subscription is active.

If you face any issues accessing your free resources, contact us at `customercare@packt.com`.

Purchased from Amazon and Other Sources

If you've purchased from sources other than the ones mentioned above (like *Amazon*), you'll need to unlock the resources first by entering your unique sign-up code provided in this section. **Unlocking takes less than 10 minutes, can be done from any device, and needs to be done only once**. Follow these five easy steps to complete the process:

STEP 1

Open the link `https://packt.link/DP900UnlockResources` OR scan the following **QR code** (*Figure 9.1*):

Figure 9.1: QR code for the page that lets you unlock this book's free online content

Either of those links will lead to the following page as shown in *Figure 9.2*:

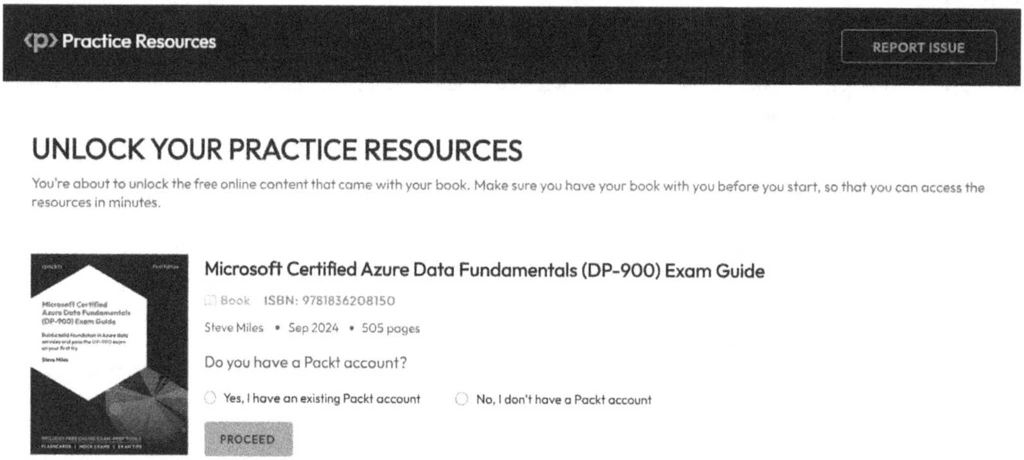

Figure 9.2: Unlock page for the online practice resources

STEP 2

If you already have a Packt account, select the option `Yes, I have an existing Packt account`. If not, select the option `No, I don't have a Packt account`.

If you don't have a Packt account, you'll be prompted to create a new account on the next page. It's free and only takes a minute to create.

Click `Proceed` after selecting one of those options.

STEP 3

After you've created your account or logged in to an existing one, you'll be directed to the following page as shown in *Figure 9.3*.

Make a note of your unique unlock code:

```
NHT9462
```

Type in or copy this code into the text box labeled 'Enter Unique Code':

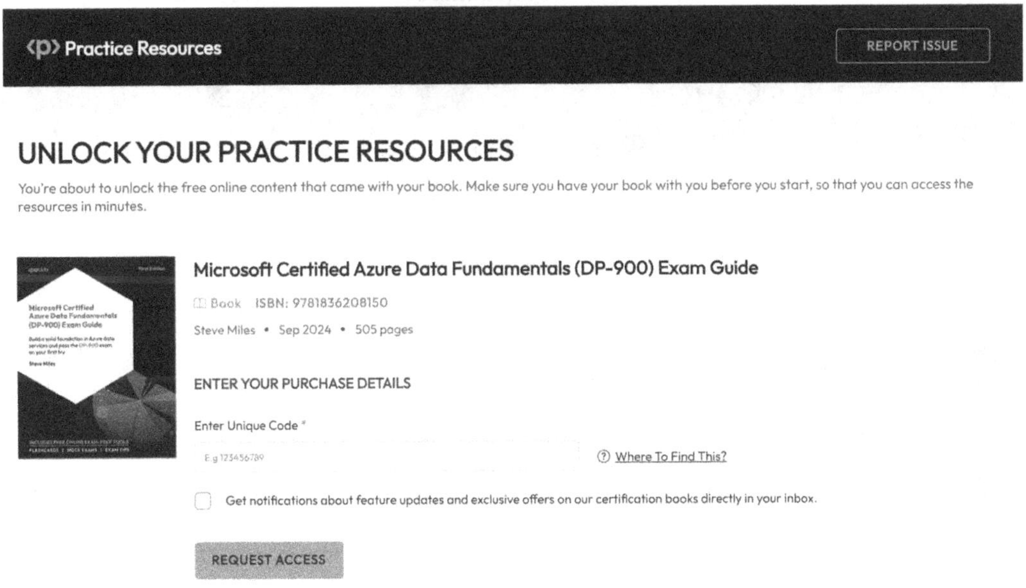

Figure 9.3: Enter your unique sign-up code to unlock the resources

> **Troubleshooting tip**
>
> After creating an account, if your connection drops off or you accidentally close the page, you can reopen the page shown in *Figure 9.2* and select `Yes, I have an existing account`. Then, sign in with the account you had created before you closed the page. You'll be redirected to the screen shown in *Figure 9.3*.

STEP 4

> **Note**
>
> You may choose to opt into emails regarding feature updates and offers on our other certification books. We don't spam, and it's easy to opt out at any time.

Click `Request Access`.

STEP 5

If the code you entered is correct, you'll see a button that says, OPEN PRACTICE RESOURCES, as shown in *Figure 9.4*:

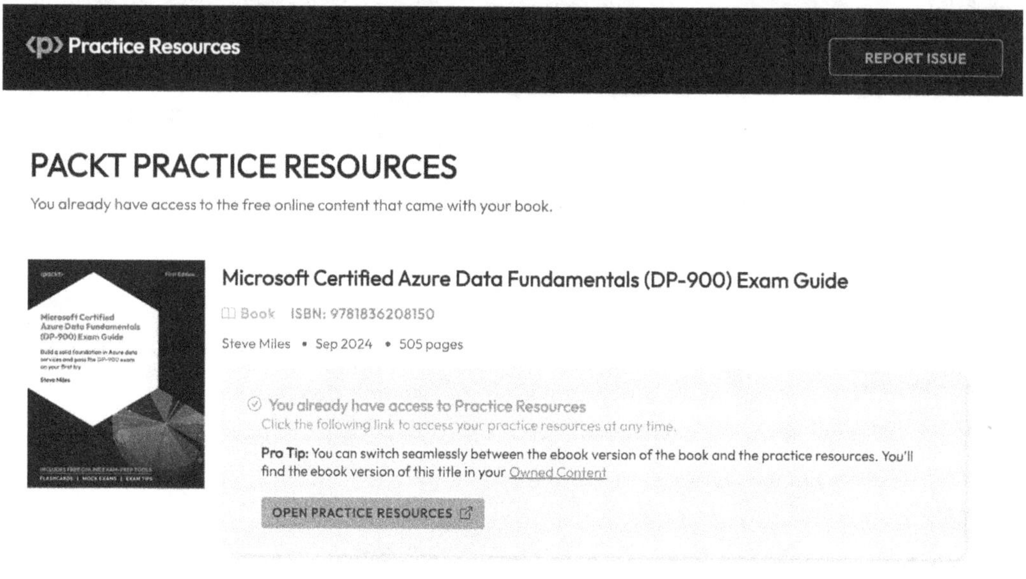

Figure 9.4: Page that shows up after a successful unlock

Click the OPEN PRACTICE RESOURCES link to start using your free online content. You'll be redirected to the Dashboard shown in *Figure 9.5*:

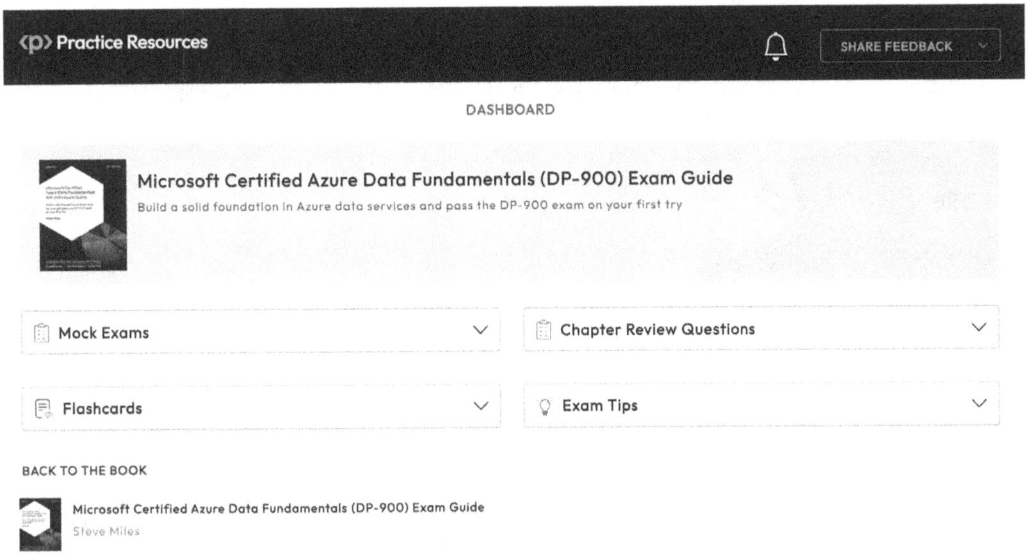

Figure 9.5: Dashboard page for DP-900 practice resources

> **Bookmark this link**
>
> Now that you've unlocked the resources, you can come back to them anytime by visiting `https://packt.link/DP900PracticeResources` or scanning the following QR code provided in *Figure 9.6*:

Figure 9.6: QR code to bookmark practice resources website

Troubleshooting Tips

If you're facing issues unlocking, here are three things you can do:

- Double-check your unique code. All unique codes in our books are case-sensitive and your code needs to match exactly as it is shown in *STEP 3*.

- If that doesn't work, use the `Report Issue` button located at the top-right corner of the page.

- If you're not able to open the unlock page at all, write to `customercare@packt.com` and mention the name of the book.

Share Feedback

If you find any issues with the platform, the book, or any of the practice materials, you can click the `Share Feedback` button from any page and reach out to us. If you have any suggestions for improvement, you can share those as well.

Back to the Book

To make switching between the book and practice resources easy, we've added a link that takes you back to the book (*Figure 9.7*). Click it to open your book in Packt's online reader. Your reading position is synced so you can jump right back to where you left off when you last opened the book.

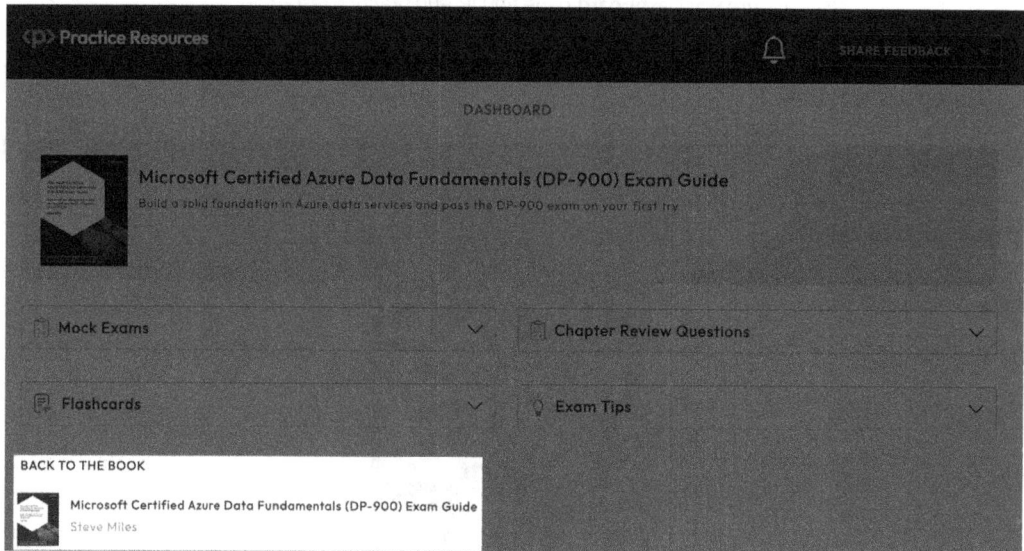

Figure 9.7: Dashboard page for DP-900 practice resources

> **Note**
>
> Certain elements of the website might change over time and thus may end up looking different from how they are represented in the screenshots of this book.

Index

X

‹packt›

Subscribe to our online digital library for full access to over 7,000 books and videos, as well as industry leading tools to help you plan your personal development and advance your career. For more information, please visit our website.

Why subscribe?

- Spend less time learning and more time coding with practical eBooks and Videos from over 4,000 industry professionals
- Improve your learning with Skill Plans built especially for you
- Get a free eBook or video every month
- Fully searchable for easy access to vital information
- Copy and paste, print, and bookmark content

At www.packtpub.com, you can also read a collection of free technical articles, sign up for a range of free newsletters, and receive exclusive discounts and offers on Packt books and eBooks.

Other Books You May Enjoy

If you enjoyed this book, you may be interested in these other books by Packt:

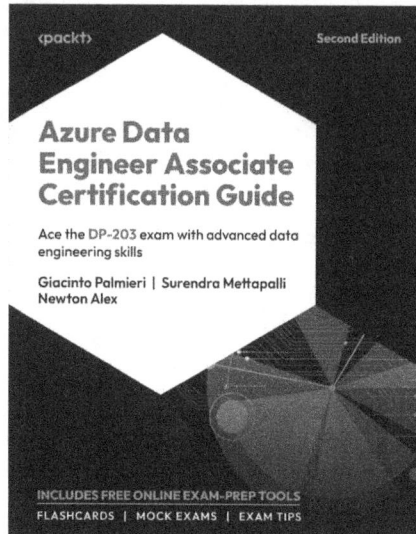

Azure Data Engineer Associate Certification Guide

Giacinto Palmieri, Mr. Surendra Mettapalli, and Newton Alex

ISBN: 978-1-80512-468-9

- Design and implement data lake solutions with batch and stream pipelines
- Secure data with masking, encryption, RBAC, and ACLs
- Perform standard extract, transform, and load (ETL) and analytics operations
- Implement different table geometries in Azure Synapse Analytics
- Write Spark code, design ADF pipelines, and handle batch and stream data
- Use Azure Databricks or Synapse Spark for data processing using Notebooks
- Leverage Synapse Analytics and Purview for comprehensive data exploration
- Confidently manage VMs, VNETS, App Services, and more

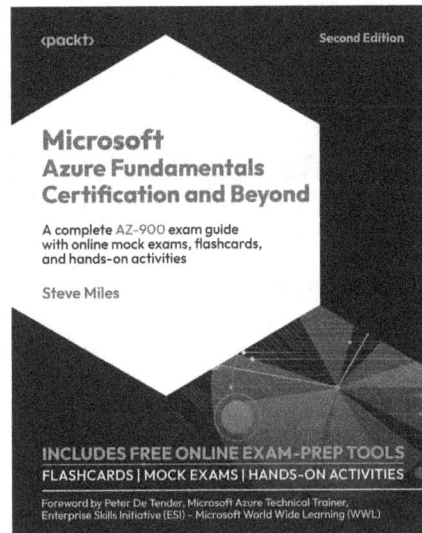

Microsoft Azure Fundamentals Certification and Beyond

Steve Miles

ISBN: 978-1-83763-059-2

- Become proficient in foundational cloud concepts
- Develop a solid understanding of core components of the Microsoft Azure cloud platform
- Get to grips with Azure's core services, deployment, and management tools
- Implement security concepts, operations, and posture management
- Explore identity, governance, and compliance features
- Gain insights into resource deployment, management, and monitoring

Share Your Thoughts

Now you've finished *Microsoft Certified Azure Data Fundamentals (DP-900) Exam Guide*, we'd love to hear your thoughts! Scan the QR code below to go straight to the Amazon review page for this book and share your feedback or leave a review on the site that you purchased it from.

https://packt.link/r/1836208154

Your review is important to us and the tech community and will help us make sure we're delivering excellent quality content.

Download a Free PDF Copy of This Book

Thanks for purchasing this book!

Do you like to read on the go but are unable to carry your print books everywhere?

Is your eBook purchase not compatible with the device of your choice?

Don't worry, now with every Packt book you get a DRM-free PDF version of that book at no cost.

Read anywhere, any place, on any device. Search, copy, and paste code from your favorite technical books directly into your application.

The perks don't stop there, you can get exclusive access to discounts, newsletters, and great free content in your inbox daily.

Follow these simple steps to get the benefits:

1. Scan the QR code or visit the link below:

https://packt.link/free-ebook/9781836208150

2. Submit your proof of purchase.
3. That's it! We'll send your free PDF and other benefits to your email directly.